# Hardy Spaces and Potential Theory on $C^1$ Domains in Riemannian Manifolds

of the
American Mathematical Society

Number 894

Hardy Spaces
and Potential Theory
on $C^1$ Domains in
Riemannian Manifolds

Martin Dindoš

January 2008 • Volume 191 • Number 894 (fourth of 5 numbers) • ISSN 0065-9266

**American Mathematical Society**
Providence, Rhode Island

2000 *Mathematics Subject Classification.* Primary 42B30, 35J55; Secondary 42B20, 58J32.

---

**Library of Congress Cataloging-in-Publication Data**

Dindoš, Martin.
　Hardy spaces and potential theory on $C^1$ domains in Riemannian manifolds / Martin Dindoš.
　　p. cm. — (Memoirs of the American Mathematical Society, ISSN 0065-9266 ; no. 894)
　"Volume 191, number 894 (fourth of 5 numbers)."
　Includes bibliographical references.
　ISBN 978-0-8218-4043-6 (alk. paper)
　1. Hardy spaces.　2. Riemannian manifolds.　3. Potential theory (Mathematics).　I. Title.
QA331.D635　2007
515'.2433—dc22
　　　　　　　　　　　　　　　　　　　　　　　　　　　　　　　　　　　　　　　　　2007060557

---

## Memoirs of the American Mathematical Society

This journal is devoted entirely to research in pure and applied mathematics.

**Subscription information.** The 2008 subscription begins with volume 191 and consists of six mailings, each containing one or more numbers. Subscription prices for 2008 are US\$675 list, US\$540 institutional member. A late charge of 10% of the subscription price will be imposed on orders received from nonmembers after January 1 of the subscription year. Subscribers outside the United States and India must pay a postage surcharge of US\$38; subscribers in India must pay a postage surcharge of US\$43. Expedited delivery to destinations in North America US\$53; elsewhere US\$130. Each number may be ordered separately; *please specify number* when ordering an individual number. For prices and titles of recently released numbers, see the New Publications sections of the *Notices of the American Mathematical Society*.

**Back number information.** For back issues see the *AMS Catalog of Publications*.

Subscriptions and orders should be addressed to the American Mathematical Society, P. O. Box 845904, Boston, MA 02284-5904, USA. *All orders must be accompanied by payment.* Other correspondence should be addressed to 201 Charles Street, Providence, RI 02904-2294, USA.

**Copying and reprinting.** Individual readers of this publication, and nonprofit libraries acting for them, are permitted to make fair use of the material, such as to copy a chapter for use in teaching or research. Permission is granted to quote brief passages from this publication in reviews, provided the customary acknowledgment of the source is given.

Republication, systematic copying, or multiple reproduction of any material in this publication is permitted only under license from the American Mathematical Society. Requests for such permission should be addressed to the Acquisitions Department, American Mathematical Society, 201 Charles Street, Providence, Rhode Island 02904-2294, USA. Requests can also be made by e-mail to reprint-permission@ams.org.

---

*Memoirs of the American Mathematical Society* is published bimonthly (each volume consisting usually of more than one number) by the American Mathematical Society at 201 Charles Street, Providence, RI 02904-2294, USA. Periodicals postage paid at Providence, RI. Postmaster: Send address changes to Memoirs, American Mathematical Society, 201 Charles Street, Providence, RI 02904-2294, USA.

© 2008 by the American Mathematical Society. All rights reserved.
Copyright of individual articles may revert to the public domain 28 years
after publication. Contact the AMS for copyright status of individual articles.
This publication is indexed in *Science Citation Index*®, *SciSearch*®, *Research Alert*®, *CompuMath Citation Index*®, *Current Contents*®/*Physical, Chemical & Earth Sciences*.
Printed in the United States of America.

∞ The paper used in this book is acid-free and falls within the guidelines
established to ensure permanence and durability.
Visit the AMS home page at http://www.ams.org/

10 9 8 7 6 5 4 3 2 1　　13 12 11 10 09 08

# Contents

| | |
|---|---:|
| Abstract | vi |
| Chapter 0. Introduction | 1 |
| Chapter 1. Background and Definitions | 4 |
| §1.1. Notation, terminology and known results | 4 |
| §1.2. Hardy spaces and layer potentials | 6 |
| Chapter 2. The Boundary Layer Potentials | 9 |
| §2.1. Compactness of operators $K$, $K^*$ | 9 |
| §2.2. Invertibility of $\pm\frac{1}{2}I + K$, $\pm\frac{1}{2}I + K^*$ | 19 |
| Chapter 3. The Dirichlet problem | 21 |
| §3.1. $L^p$ boundary data | 21 |
| §3.2. Hardy space boundary data | 23 |
| §3.3. Hölder space boundary data | 25 |
| Chapter 4. The Neumann problem | 27 |
| §4.1. $L^p$ boundary data | 27 |
| §4.2. Hardy space boundary data | 27 |
| §4.3. Hölder space boundary data | 29 |
| Chapter 5. Compactness of Layer Potentials, Part II; The Dirichlet regularity problem | 31 |
| §5.1. Preliminaries | 31 |
| §5.2. Compactness and invertibility of $K$ on Sobolev space $H^{1,p}$ | 33 |
| §5.3. Compactness and invertibility of $K$ on Hardy-Sobolev space $H^{1,p}$ | 38 |
| §5.4. Dirichlet regularity problem, Sobolev $H^{1,p}$ ($1 < p < \infty$) data | 41 |
| §5.5. Dirichlet regularity problem, $H^{1,p}$ ($(n-1)/n < p \leq 1$) data | 42 |
| Chapter 6. The equivalence of Hardy space definitions | 44 |
| §6.1. Preliminaries | 44 |
| §6.2. $C$-suharmonicity | 45 |
| §6.3. The main step | 47 |
| §6.4. The equivalence theorem on $C^1$ domains | 50 |
| §6.5. The equivalence theorem on Lipschitz domains | 51 |
| APPENDIX A. Variable Coefficient Cauchy Integrals | 53 |
| APPENDIX B. One Result on the Maximal Operator | 65 |
| Bibliography | 77 |

# Abstract

We study Hardy spaces on $C^1$ and Lipschitz domains in Riemannian manifolds. Hardy spaces, originally introduced in 1920 in complex analysis setting, are invaluable tool in harmonic analysis. For this reason these spaces have been studied extensively by many authors.

Our main result is an equivalence theorem proving that the definition of Hardy spaces by conjugate harmonic functions is equivalent to the atomic definition of these spaces. We establish this theorem in any dimension if the domain is $C^1$, in case of a Lipschitz domain the result holds if dim $M \leq 3$. The remaining cases for Lipschitz domains remain open. This result is a nontrivial generalization of flat ($\mathbb{R}^n$) equivalence theorems due to Fefferman, Stein, Dahlberg and others.

The material presented here required to develop potential theory approach for $C^1$ domains on Riemannian manifolds in the spirit of earlier works by Fabes, Jodeit and Rivière and recent results by Mitrea and Taylor. In particular, the first part of this work is of interest in itself, since we consider the boundary value problems for the Laplace-Beltrami operator. We prove that both Dirichlet and Neumann problem for Laplace-Beltrami equation are solvable for any given boundary data in $L^p(\partial\Omega)$, where $1 < p < \infty$. Same remains true in Hardy spaces $\hbar^p(\partial\Omega)$ for $(n-1)/n < p \leq 1$.

In the whole work we work with Riemannian metric $g$ with smallest possible regularity. In particular, mentioned results for the Laplace-Beltrami equation require Hölder class regularity of the metric tensor; the equivalence theorem requires $g$ in $C^{1,1}$.

---

Received by the editor April 30, 2001; and in revised form September 21, 2004.

2000 *Mathematics Subject Classification.* Primary 42B30, 35J55; Secondary 42B20, 58J32.

*Key words and phrases.* Hardy spaces, Potential theory, Riemannian manifolds, rough domains, Laplace-Beltrami equation, boundary value problems.

CHAPTER 0

# Introduction

The concept of Hardy spaces arose between 1910-20 in the context of Fourier series and complex analysis in one variable. The important feature of the Hardy spaces is that they naturally extend the interpolation scale of the $L^p$ spaces, $1 < p < \infty$ to all $p > 0$. Classically, the theory works only in one complex variable, as the attempts to extend the Hardy spaces to several complex variables ran into a lot of trouble. For a long time this had blocked all attempts to extend the deeper properties of Hardy spaces to several variables.

Around 1960 Elias Stein and Guido Weiss [28] realized that the several complex variables was a narrow generalization of Hardy spaces for the purposes of Fourier analysis. Their idea was very simple, yet completely changed the subject. They realized that the important feature of an analytic function of one complex variable is that both its real and imaginary parts are harmonic functions. In several variables, the gradient of a harmonic function is a system $(u_1, u_2, \ldots, u_n)$ of functions on $\mathbb{R}^n$ that satisfies the Stein-Weiss Cauchy-Riemann equations

$$\frac{\partial u_j}{\partial x_k} = \frac{\partial u_k}{\partial x_j}, \qquad \sum_k \frac{\partial u_k}{\partial x_k} = 0.$$

This idea led to the development of Hardy spaces on $\mathbb{R}^n$. Hardy spaces attracted the interest of many mathematicians and become a centerpiece of harmonic analysis. The reason for this success is clear, these spaces are extremely useful tool for studying various problems in harmonic analysis, partial differential equations, and probability.

The 'hallmark' property of Hardy spaces is the multiplicity of definitions. On $\mathbb{R}^n$ several definitions of Hardy spaces can be given which are equivalent and therefore can be used interchangeably. It was Charles Fefferman and Elias Stein who first discovered this feature in [17] and used it to prove that singular integral operators are bounded on Hardy spaces for all $p > 0$.

The usefulness of having several equivalent definitions cannot be overemphasized. Different mathematical problems requires different approaches and the fact that one can use the most convenient definition in the appropriate setting makes work much easier.

In particular, as presented in [26], the Hardy spaces $\hbar^p(\mathbb{R}^n)$ can be defined in terms of *atoms*; via harmonic functions in the upper half space $\mathbb{R}^{n+1}_+$ using a certain *maximal function* and finally, for $(n-1)/n < p \leq 1$ also using the *conjugate harmonic functions* in $\mathbb{R}^{n+1}_+$.

The success of Hardy spaces on $\mathbb{R}^n$ led to several successful attempts to generalize this concept further. For example, Fabes and Kenig in [15] introduced the Hardy space $\hbar^1$ on a boundary of a $C^1$ domain in $\mathbb{R}^n$ and established an equivalence theorem in this setting. Namely, they proved the equivalence between atomic and maximal function definitions.

The settings where Hardy spaces can be defined can be extended further. Namely given any $n$-dimensional compact Riemannian manifold $M$ and an open subset $\Omega$ on $M$ with Lipschitz boundary, we can without complication modify the atomic definition of Hardy spaces from $\mathbb{R}^n$ and define the Hardy space $\hbar^p$ on the boundary of the set $\Omega$ (cf. Chapter 1 of this work).

The natural and important question arises - whether the other two characterizations of the Hardy space can also be modified and whether they yield the same space.

The main goal of this work is to provide an answer to this question. Namely, in the main result we show that the definition using conjugate harmonic functions adapted to our setting is equivalent to the atomic definition provided the metric tensor on $M$ is of class $C^{1,1}$ and the boundary $\partial\Omega$ is $C^1$. If dim $M \leq 3$ we have this result even for Lipschitz domains.

This result for the standard Laplacian on $\mathbb{R}^n$ was obtained by Dahlberg in [7]. He also asked what happen for $n \geq 4$ for Lipschitz domains (cf. problem #2 in [7]). This question remains open.

At this point it is important to say that our results on manifolds are not a straightforward generalization of the flat $\mathbb{R}^n$ case. Two major obstacles hamper such approach.

The first problem is the basic nature of a Riemannian manifold - the existence of a curvature. In contrast, the flatness of $\mathbb{R}^n$ give rise to important features of harmonic functions which cannot be observed in the presence of nonzero curvature. In particular, for a harmonic function $u$ on $\mathbb{R}^n$, the function $|\nabla u|^q$ is subharmonic (i.e. $\Delta(|\nabla u|^q) \geq 0$) for any $q \geq (n-1)/n$. It is straightforward to check that this is not true for harmonic functions on manifolds. Hence a different approach has to be developed, using weaker notions transferable to manifolds.

The second obstacle was to develop further the techniques of boundary layer potentials on $C^1$ domains in Riemannian manifolds.

The boundary layer techniques have been successful in the treatment of the Laplace equation and certain other constant coefficient elliptic partial differential equations on $C^1$ and Lipschitz domains in Euclidean space. The first successful result in this direction for the Dirichlet problem for the classical flat Laplacian was in the paper by Dahlberg [6]. The idea of the paper was to use certain estimates on the harmonic measure. The breakthrough for the $C^1$ domains followed in the paper by Fabes, Jodeit and Rivère [14] who showed the solvability of the Laplace boundary problem in $L^p$, $1 < p < \infty$ for both Neumann and Dirichlet problem, as well as the Dirichlet regularity problem, i.e., with boundary data in the Sobolev space $H^{1,p}$.

Finally, the question whether similar results can be obtained for a Lipschitz boundary was affirmatively answered for flat Laplacian and restricted range of $p$ by Dahlberg and Kenig in [10] and Verchota in [30]. Related questions for the system of elastostatics on Lipschitz domains were considered by Dahlberg, Kenig and Verchota in [11]; the Stokes system on Lipschitz domains is considered in [16] by Fabes, Kenig and Verchota. Of interest is also work by M. Mitrea, D. Mitrea and Pipher [20] on vector potential theory on nonsmooth domains in $\mathbb{R}^3$.

Recently, a very successful approach generalizing these techniques to the Laplace-Beltrami equation on Riemannian manifolds for Lipschitz domains has appeared in [**22**], [**23**] and [**24**] by M. Mitrea and M. Taylor. In [**25**] the authors also stated several results about the solvability of the Laplace-Beltrami equation in $L^p$, $1 < p < \infty$ and in the Besov spaces for $\Omega$ with $C^1$ boundary.

In this work we develop approach for $C^1$ domains similar to the work [**14**] by Fabes, Jodeit and Riviere. We are particularly interested in the optimal range of $p$ for which solvability and regularity of the solution of Laplace-Beltrami equation in $L^p$ and Hardy spaces $\hbar^p$ with Dirichlet and Neumann boundary data can be established. For the sake of completeness of the exposition we also state the results (obtained by different techniques) from [**25**] for $L^p$, $1 < p < \infty$. The results for the Hardy spaces are new.

Let us mention one somewhat unexpected impact of this work. In Appendix B we introduce a Banach space $\mathcal{D}^{0,p}$. We originally introduced this space mainly for technical reasons to prove one particular result necessary for the main exposition. However, as we discovered in our subsequent works [**12**], [**13**], which consider semilinear elliptic problem for Lipschitz domains on Riemannian manifolds, the usefulness of this space does not end there. In some instances it can be successfully used instead of the Sobolev space $H^{s,p}(\Omega)$, when the use of Sobolev space is inconvenient or impossible.

To bring this introduction to an end, we briefly describe the organization of this paper. In Chapter 1 we give basic definitions and state results necessary for our work that can be found in the literature, Chapters 2 and 5 are devoted to the question of compactness of layer potential on $C^1$ domains. In Chapters 3, 4 and 5 we present solvability of the Dirichlet, Neumann and Dirichet regularity problem for $C^1$ domains in Riemannian manifolds, respectively. These results are of interest in themselves and therefore we spend considerable space to present them. Finally, the main result on equivalence of definitions of the Hardy space is presented in Chapter 6. The work also has two appendices containing material outside the main flow of the exposition. Appendix A presents various results on variable coefficient Cauchy integrals and Appendix B contains material on the actions of the operator $(\Delta - V)^{-1}$ on functions from the space $\mathcal{D}^{0,p}$ already mentioned above.

**Acknowledgments.** Presented material is an outgrowth of my Ph.D. dissertation that I wrote as a student of Michael Taylor at the University of North Carolina at Chapel Hill. I am very grateful to him for his guidance and support to me while working on my thesis. His knowledge and deep insight helped me to sharpen my arguments and avoid many embarrassing mistakes.

I have had the good fortune to benefit from conversations with Alan McIntosh from Australian National University and Marius Mitrea from University of Missouri at Columbia. I want to thank them for answering my emails with questions and helping me to overcome certain obstacles I came across. Their contribution is acknowledged at appropriate places.

Finally, I would also like to thank the anonymous referee and the editor William Beckner for their valuable suggestions that improved overall organization of this paper.

CHAPTER 1

# Background and Definitions

## 1.1. Notation, terminology and known results

We recall the setting of the papers [**22**], [**23**] and [**24**], which is used throughout this work. Let $M$ be a smooth, compact Riemannian manifold, of real dimension $\dim M = n$, with a Riemannian metric tensor, which is assumed to be of Hölder class $C^\alpha$. (For some results we will have to assume higher regularity of the metric tensor.) That is, $M$ is covered by local coordinate charts with the components $g_{jk}$ of the metric tensor being of Hölder class $C^\alpha$. The Laplace-Beltrami operator on $M$ is defined by

$$(1.1) \qquad \Delta : H^{1,p}(M) \to H^{-1,p}(M), \quad (\Delta u, v) \stackrel{\text{def}}{=} -\int_M \langle du, dv \rangle \, d\text{Vol},$$

where $H^{s,p}$, $s \in \mathbb{R}$, $p \in (1, \infty)$, denotes the usual class of $L^p$ Sobolev spaces on $M$. Here, the metric tensor determines the pointwise inner product on $T_x^* M$ and the volume element $d\text{Vol}$. In local coordinates

$$(1.2) \qquad \Delta u = g^{-1/2} \partial_j (g^{jk} g^{1/2} \partial_k u).$$

Here we use the summation convention, take $(g^{jk})$ to be inverse matrix to $(g_{jk})$ and set $g = \det(g_{jk})$. For $V \in L^\infty(M)$ we introduce the second order, elliptic differential operator

$$(1.3) \qquad L = \Delta - V.$$

We assume $V \geq 0$ on $M$ and also $V > 0$ on a set of positive measure in each connected component of $M \setminus \overline{\Omega}$. Here $\Omega \subset M$ is assumed to be open, connected and with $C^1$ boundary.

Our goal is to use boundary layer methods in the treatment of the Dirichlet boundary problem

$$(1.4) \qquad Lu = 0 \text{ in } \Omega, \quad u\big|_{\partial \Omega} = f \in L^p(\partial \Omega),$$

and the Neumann boundary problem

$$(1.5) \qquad Lu = 0 \text{ in } \Omega, \quad \partial_\nu u\big|_{\partial \Omega} = g \in L^p(\partial \Omega).$$

Here $\partial_\nu = \partial/\partial_\nu$ is the normal derivative on $\partial \Omega$. Here and hereafter, all boundary traces are taken in the nontangential limit sense, i.e., given a function $u$ defined and continuous on $\Omega$, set

$$(1.6) \qquad u\big|_{\partial \Omega}(x) \stackrel{\text{def}}{=} \lim_{\substack{y \to x \\ y \in \gamma(x)}} u(y), \qquad x \in \partial \Omega,$$

when this limit exists. In (1.6) $\gamma(x) \subset \Omega$ is a nontangential approach region with "vertex" at $x$; cf. [**22**] for more details. Furthermore,

$$\partial_\nu u\big|_{\partial\Omega}(x) \stackrel{\text{def}}{=} \langle \nu, du\big|_{\partial\Omega}\rangle, \tag{1.7}$$

where $\nu \in T^*(M)$ is the (outward) unit normal to $\partial\Omega$.

Now using the approach in [**22**] under our hypothesis on $V$, the operator

$$L : H^{-1,p}(M) \to H^{1,p}(M) \tag{1.8}$$

is an isomorphism, for each $p \in (1, \infty)$. Denote by $E(x, y)$ the integral kernel of $L^{-1}$, so

$$L^{-1}u(x) = \int_M E(x,y)u(y)\, d\text{Vol}(y), \qquad x \in M. \tag{1.9}$$

For a function $f : \partial\Omega \to \mathbb{R}$ define the single layer potential

$$\mathcal{S}f(x) = \int_{\partial\Omega} E(x,y)f(y)\, d\sigma(y), \qquad x \notin \partial\Omega, \tag{1.10}$$

where $d\sigma$ is the natural area element on $\partial\Omega$. Similarly, we define the double layer potential by

$$\mathcal{D}f(x) = \int_{\partial\Omega} \frac{\partial E}{\partial \nu_y}(x,y)f(y)\, d\sigma(y), \qquad x \notin \partial\Omega. \tag{1.11}$$

The following results on the behavior of these potentials were demonstrated in [**22**], extending previously known results for the flat Euclidean case.

Define $\Omega_+ = \Omega$ and $\Omega_- = M \setminus \overline{\Omega}$, note that $\Omega_\pm$ are $C^1$ domains. Given $f \in L^p(\partial\Omega)$, $1 < p < \infty$ we have, for a.e. $x \in \partial\Omega$,

$$\mathcal{S}f\big|_{\partial\Omega_+}(x) = \mathcal{S}f\big|_{\partial\Omega_-}(x) = Sf(x), \tag{1.12}$$

and

$$\mathcal{D}f(x)\big|_{\partial\Omega_\pm}(x) = (\pm\tfrac{1}{2}I + K)f(x), \tag{1.13}$$

where for a.e. $x \in \partial\Omega$

$$\begin{aligned}Sf(x) &= \int_{\partial\Omega} E(x,y)f(y)\, d\sigma(y), \\ Kf(x) &= \text{P.V.} \int_{\partial\Omega} \frac{\partial E}{\partial \nu_y}(x,y)f(y)\, d\sigma(y).\end{aligned} \tag{1.14}$$

Here P.V.$\int_{\partial\Omega}$ indicates that the integral is taken in the principal value sense. More concretely for a fixed smooth background metric which induces a distance function on $M$ we can talk about balls on $\partial\Omega$. P.V.$\int_{\partial\Omega}$ is defined in the sense of removing such small geodesic ball around the point $x \in \partial\Omega$ and then passing to the limit. Furthermore for a.e. $x \in \partial\Omega$

$$\partial_\nu \mathcal{S}f\big|_{\partial\Omega_\pm}(x) = (\mp\tfrac{1}{2}I + K^*)f(x), \tag{1.15}$$

where $K^*$ is the formal adjoint of $K$. Moreover, the operators

$$K, K^* : L^p(\partial\Omega) \to L^p(\partial\Omega), \qquad 1 < p < \infty, \tag{1.16}$$

and

(1.17) $$S: L^p(\partial\Omega) \to H^{1,p}(\partial\Omega), \qquad 1 < p < \infty,$$

are bounded and we have nontangential maximal function estimates

(1.18) $$\|(\nabla \mathcal{S}f)^*\|_{L^p(\partial\Omega)} \leq C_p \|f\|_{L^p(\partial\Omega)}, \quad \|(\mathcal{D}f)^*\|_{L^p(\partial\Omega)} \leq C_p \|f\|_{L^p(\partial\Omega)},$$

for $1 < p < \infty$. Here and hereafter, if $u$ is defined on $\Omega$ then $u^*$ will denote the nontangential maximal function of $u$, defined at the boundary points by

(1.19) $$u^*(x) = \sup\{|u(y)| : y \in \gamma(x)\}, \qquad x \in \partial\Omega.$$

The major difference between the treatment in the Lipschitz and the $C^1$ case is that in the $C^1$ case operators $K, K^* : L^p(\partial\Omega) \to L^p(\partial\Omega)$ are compact for $p \in (1, \infty)$ and hence the operators

(1.20) $$\pm \tfrac{1}{2}I + K, \pm \tfrac{1}{2}I + K^* : \ L^p(\partial\Omega) \to L^p(\partial\Omega)$$

are Fredholm in this range of $p$. In the Lipschitz case establishing the Fredholmness of these operators is a major obstacle which reduces the range of $p$ in which solvability of (1.4) and (1.5) can be established.

In order to establish compactness of the operators $K, K^*$ we need to understand the structure of the singularity in the kernel $E(x, y)$ of $L^{-1}$ along the diagonal. The main result of [**24**] in this direction is the decomposition of this kernel as

(1.21) $$E(x, y) = g(y)^{-1/2} \{e_0(x - y, y) + e_1(x, y)\},$$

where:

(1.22) $$e_0(x - y, y) = C_n \left( \sum g_{jk}(y)(x_j - y_j)(x_k - y_k) \right)^{-(n-2)/2}.$$

Here $C_n$ is a suitable constant and the residual term $e_1(x, y)$ satisfies

(1.23) $$\begin{aligned} |e_1(x, y)| &\leq C_\varepsilon |x - y|^{-(n-2-\alpha+\varepsilon)}, \\ |\nabla_x e_1(x, y)| &\leq C_\varepsilon |x - y|^{-(n-1-\alpha+\varepsilon)}, \end{aligned}$$

where $\alpha$ is the Hölder coefficient of continuity of the metric tensor $g$ and $\varepsilon > 0$.

## 1.2. Hardy spaces and layer potentials

To bring this chapter to an end, we discuss several definitions which are of importance to us. By $\hbar^p(\partial\Omega)$ we mean the localization of the atomic Hardy space $\hbar^p_{\mathrm{at}}(\partial\Omega)$. We recall that a function $f \in L^\infty(\partial\Omega)$ is an $p$-atom for $(n-1)/n < p \leq 1$ if

(1.24) $$\operatorname{supp} f \subset B_r(x_0) \cap \partial\Omega$$

for some $x_0 \in \partial\Omega$, $r \in (0, \operatorname{diam} \Omega]$, and

(1.25) $$\|f\|_{L^\infty(\partial\Omega)} \leq \frac{1}{r^{(n-1)/p}}, \qquad \int_{\partial\Omega} f \, d\sigma = 0.$$

Then $g \in L^p(\partial\Omega)$ is said to belong to $\hbar^p_{\mathrm{at}}(\partial\Omega)$ provided it can be written in the form

(1.26) $$g = \sum_{\nu \geq 1} a_\nu f_\nu, \quad f_\nu \text{ an } p\text{-atom}, \quad \sum_{\nu \geq 1} |a_\nu|^p < \infty.$$

## 1.2. HARDY SPACES AND LAYER POTENTIALS

There is a "norm" defined by

$$\|g\|_{\hbar^p_{\mathrm{at}}(\partial\Omega)} = \inf\left\{ (\sum_{\nu\geq 1} |a_\nu|^p)^{1/p} : g = \sum_{\nu\geq 1} a_\nu f_\nu,\ f_\nu \text{ a } p\text{-atom}\right\}. \tag{1.27}$$

Then we can set

$$\hbar^p(\partial\Omega) \stackrel{\text{def}}{=} \hbar^p_{\mathrm{at}}(\partial\Omega) + \mathcal{C} = \hbar^p_{\mathrm{at}}(\partial\Omega) + L^q(\partial\Omega), \quad \forall q \in (1,\infty], \tag{1.28}$$

where $\mathcal{C}$ consists of functions on $\partial\Omega$ that are constant on each connected component of $\partial\Omega$. Under $f \mapsto \varphi f$ this space is a module over $C^r(\partial\Omega)$, for any $r > (n-1)(p^{-1}-1)$. If $(n-1)/n < p < 1$ then $\hbar^p(\partial\Omega)$ is only a quasi-Banach space and its dual is

$$(\hbar^p(\partial\Omega))^* = C^\alpha(\partial\Omega) \qquad \text{for } \alpha = (n-1)(p^{-1}-1). \tag{1.29}$$

As usual, by $C^\alpha$ we denoted the space of Hölder continuous functions on $\partial\Omega$. The dual of the Banach space $\hbar^1(\partial\Omega)$ is $\mathrm{bmo}(\partial\Omega)$.

Now we can briefly discuss layer potentials on $\hbar^p(\partial\Omega)$, $(n-1)/n < p \leq 1$ following the supplement B of [**23**]. Using the decomposition (1.21) of the kernel $E(x,y)$ we can apply Proposition A.8. of the appendix on the $\nabla_x e_0(x-y,y)$, provided the Hölder coefficient $\alpha$ is bigger than $(n-1)(p^{-1}-1)$. Also $e_1(x,y)$ satisfies (1.23) hence using analysis similar to (B.5)-(B.6) in [**23**] for this term we can show:

PROPOSITION 1.1. *For* $\max\{\frac{n-1}{n}, \frac{n-1}{\alpha+n-1}\} < p \leq 1$

$$\|(\nabla \mathcal{S}f)^*\|_{L^p(\partial\Omega)} \leq C\|f\|_{\hbar^p(\partial\Omega)}, \tag{1.30}$$

*uniformly for* $f \in \hbar^p(\partial\Omega)$. *In particular if the metric tensor on $M$ is Lipschitz (1.30) holds for* $(n-1)/n < p \leq 1$.

Similarly according Proposition B.6 of [**23**] in our setting we can prove:

PROPOSITION 1.2. *The operators*

$$K^* : \hbar^p(\partial\Omega) \to \hbar^p(\partial\Omega), \tag{1.31}$$

*and*

$$\nu \wedge d\mathcal{S} : \hbar^p(\partial\Omega) \to \hbar^p(\partial\Omega), \tag{1.32}$$

*are well defined and bounded for each* $\max\{\frac{n-1}{n}, \frac{n-1}{\alpha+n-1}\} < p \leq 1$.

Here we use notation $(\nu \wedge d\mathcal{S})f$ instead of $\partial_\nu \mathcal{S}f$ since for $p < 1$ $f \in \hbar^p(\partial\Omega)$ is in general only distribution. Now we treat double layer potential. We can establish a weaker statement:

PROPOSITION 1.3. *Assume that the boundary $\partial\Omega$ is of class $C^{1+\alpha}$. Then the operators*

$$\mathcal{D}\big|_{\partial\Omega_\pm} : \hbar^p(\partial\Omega) \to \hbar^p(\partial\Omega), \tag{1.33}$$

*and*

$$K : \hbar^p(\partial\Omega) \to \hbar^p(\partial\Omega), \tag{1.34}$$

*are well defined and bounded for each* $\max\{\frac{n-1}{n}, \frac{n-1}{\alpha+n-1}\} < p \leq 1$.

PROOF. Using the decomposition (1.21) and symmetry of $E(x,y)$ we can write

$$(1.35) \qquad \frac{\partial E}{\partial \nu_y}(x,y) = \frac{1}{\sqrt{g(x)}}\left\{\frac{\partial}{\partial \nu_y}e_0(x-y,x) + \frac{\partial}{\partial \nu_y}e_1(y,x)\right\}.$$

The reason we need smoothness $C^{1+\alpha}$ of the boundary is that evaluating $\frac{\partial}{\partial \nu_y}e_0(x-y,x)$ yields

$$(1.36) \qquad \frac{\partial}{\partial \nu_y}e_0(x-y,x) = -\sum_{i=1}^{n}\frac{\partial}{\partial y_i}e_0(x-y,x)\nu^i(y),$$

where $\nu^i(y)$ is the $i$-th component of the outer normal to the boundary $\partial\Omega$ at $y$. Hence assuming only $C^1$ boundary yields that $\nu^i(y)$ is just a continuous function. However, $\hbar^p(\partial\Omega)$ is not a module over continuous functions which means that the product $\nu^i(y)f(y)$ might not be in $\hbar^p(\partial\Omega)$ for $f \in \hbar^p(\partial\Omega)$. So we really need the assumption $\partial\Omega \in C^{1+\alpha}$ and then everything works for $\max\{\frac{n-1}{n}, \frac{n-1}{\alpha+n-1}\} < p \leq 1$. Same is true for the other term in (1.35).

Now we can use Proposition A.7 on the first term in (1.35) and analysis similar to (B.5)-(B.6) on the second term to conclude:

$$(1.37) \qquad \|(\mathcal{D}f(x))^*\|_{L^p(\partial\Omega)} \leq C\|f\|_{\hbar^p(\partial\Omega)}.$$

Consider now $f$ in $L^2(\partial\Omega)$, say. Then by the approach developed in [31] in $\mathbb{R}^n$ but working also for Lipschitz domains on manifolds since it uses only the ordinary nontangential maximal operator together with cancellations based on integration by parts we get

$$(1.38) \qquad \mathcal{D}f(x) \in \hbar^p(\partial\Omega) \quad \text{and} \quad \|\mathcal{D}f(x)|_{\partial\Omega_\pm}\|_{\hbar^p(\partial\Omega)} \leq C\|(\mathcal{D}f(x))^*\|_{L^p(\partial\Omega)}.$$

Combining (1.37), (1.38) and (1.13) we have

$$(1.39) \qquad \|\mathcal{D}f(x)|_{\partial\Omega_\pm}\|_{\hbar^p(\partial\Omega)} = \|(\pm\tfrac{1}{2}I + K)f(x)\|_{\hbar^p(\partial\Omega)} \leq C\|f\|_{\hbar^p(\partial\Omega)}.$$

From this (1.34) follows by a density argument.

CHAPTER 2

# The Boundary Layer Potentials

## 2.1 Compactness of operators $K$, $K^*$

The main goal of this section is to show that the operators $K, K^*$ defined in the first chapter are compact on $L^p$ for $1 < p < \infty$. We will also establish similar result for Hardy spaces $\hbar^p$. In the second part, using this result, we will establish invertibility of the operators

(2.1) $$\pm \tfrac{1}{2} I + K, \pm \tfrac{1}{2} I + K^* : L^p(\partial \Omega) \to L^p(\partial \Omega).$$

Recall the definition of $K$ and $K^*$. We have

(2.2) $$Kf(x) = \lim_{\varepsilon \to 0+} \int_{y \in \partial\Omega, r(x,y) > \varepsilon} \frac{\partial E}{\partial \nu_y}(x,y) f(y) \, d\sigma(y),$$
$$K^* f(x) = \lim_{\varepsilon \to 0+} \int_{y \in \partial\Omega, r(x,y) > \varepsilon} \frac{\partial E}{\partial \nu_x}(x,y) f(y) \, d\sigma(y),$$

where $r(x,y)$ stands for the geodesic distance between $x, y \in M$.

To prove the result we use the following idea. We approximate the set $\Omega$ by a increasing sequence of open sets $\Omega_k$, such that $\Omega_k \nearrow \Omega$, each $\Omega_k$ has smooth boundary, and for each point $x \in \partial \Omega$ there exists a small neighborhood $U$ of $x$, such that in this neighborhood there are smooth local coordinates in which

(2.3) $$U \cap \Omega = \{x = (x', x_n) \in U : x_n > \varphi(x')\},$$
$$U \cap \Omega_k = \{x = (x', x_n) \in U : x_n > \varphi_k(x')\}.$$

Here $\varphi : \mathbb{R}^{n-1} \to \mathbb{R}$ is a $C^1$ function and $\varphi_k : \mathbb{R}^{n-1} \to \mathbb{R}$ are $C^\infty$ functions such that

(2.4) $$\varphi_k \to \varphi \quad \text{and} \quad \nabla \varphi_k \to \nabla \varphi.$$

Now we define operators $K_k$, $K_k^*$ exactly as $K$, $K^*$ are defined in (2.2); the difference is that we integrate over $\partial \Omega_k$. We will show that these operators are compact on $L^p(\partial \Omega_k)$ and $\hbar^p(\partial \Omega_k)$ and converge in the norm to $K$, $K^*$, respectively.

In order to simplify the whole thing we decompose $K$, $K^*$, $K_k$, $K_k^*$ using a partition of unity $(\phi^i)$ on $M$. We get that $K$ can be written as a sum of

(2.5) $$K = \sum_{i,j} K^{ij}, \quad \text{where:} \quad K^{ij} f(x) = \lim_{\varepsilon \to 0+} \int_{y \in O_\varepsilon} \phi^i(x) \phi^j(y) \frac{\partial E}{\partial \nu_y}(x,y) f(y) \, d\sigma(y),$$

and

(2.6) $$\mathcal{O}_\varepsilon = \{(x,y) \in \partial\Omega \times \partial\Omega : r(x,y) > \varepsilon\}.$$

The other operators are decomposed similarly.

Clearly, proving compactness of each $K^{ij}$ would suffice. As we can see immediately, if $U_i \cap U_j = \emptyset$ (where $U_i = \text{supp } \phi^i$), then the compactness of such operator is trivial, in fact it follows from the fact that the kernel of $K^{ij}$ is not singular. Also showing that $K_k^{ij} \to K^{ij}$ in the norm as $k \to \infty$ is trivial. The problematic cases are when $U_i \cap U_j \neq \emptyset$, since the operator contains singularity that has to be taken care of.

We can also assume that the partition of unity $(\phi^i)$ we picked has the property that on each $U^i \cup U^j$ (for all pairs $i,j$ for which $U_i \cap U_j \neq \emptyset$), the sets $\Omega$, $\Omega_k$ can be written as in (2.3) in some smooth coordinate chart.

Recall now the decomposition of the kernel $E(x,y)$ in (1.35). Using it we can write

(2.7)
$$K^{ij} f(x) = \lim_{\varepsilon \to 0+} \int_{y \in O_\varepsilon} \phi^i(x)\phi^j(y) \frac{1}{\sqrt{g(x)}} \frac{\partial e_0}{\partial \nu_y}(x-y, x) f(y) \, d\sigma(y)$$
$$+ \lim_{\varepsilon \to 0+} \int_{y \in O_\varepsilon} \phi^i(x)\phi^j(y) \frac{1}{\sqrt{g(x)}} \frac{\partial e_1}{\partial \nu_y}(y, x) f(y) \, d\sigma(y)$$
$$= K_1^{ij} f(x) + K_2^{ij} f(x),$$

$$K^{*ij} f(x) = \lim_{\varepsilon \to 0+} \int_{y \in O_\varepsilon} \phi^i(x)\phi^j(y) \frac{1}{\sqrt{g(y)}} \frac{\partial e_0}{\partial \nu_x}(x-y, y) f(y) \, d\sigma(y)$$
$$+ \lim_{\varepsilon \to 0+} \int_{y \in O_\varepsilon} \phi^i(x)\phi^j(y) \frac{1}{\sqrt{g(y)}} \frac{\partial e_1}{\partial \nu_x}(x, y) f(y) \, d\sigma(y)$$
$$= K_1^{*ij} f(x) + K_2^{*ij} f(x).$$

Similarly, we can decompose $K_k^{ij}$ and $K_k^{*ij}$ for any $k = 1, 2, \ldots$. The key point here is that the decomposition (1.35) of the kernel depends on chosen coordinates. This is a very desirable property, because it will allow us to pick coordinates in which the more singular piece $K_{k,1}^{*ij}$ is arbitrary small for any $k \geq 1$. Hence if we prove compactness of the other piece $K_{k,2}^{*ij}$, we get that the operator $K_k^{*ij}$ is compact for any $k \geq 1$.

We deal with these matters now. We begin with $K_{k,1}^{*ij}$. For the sake of simplicity we drop indices $i, j$ for all operators and instead of $\phi^i(x)\phi^j(y)$ we will write $\psi(x,y)$. Assume first that the metric tensor $g$ on $M$ is *smooth*. This gives us that for a fixed $k \in \mathbb{N}$ we can pick *smooth* coordinates on some small neighborhood $U$ of a point $z \in \partial \Omega_k$ such that the following holds:

(2.8)
$$\Omega_k \cap U = \{x \in \mathbb{R}^n : x_n > 0\} \cap U,$$
$$\partial \Omega_k \cap U = \{x \in \mathbb{R}^n : x_n = 0\} \cap U,$$
$$g_{jn}(x) = \delta_{jn} \qquad \text{for } x \in \partial \Omega_k,$$
$$\nu(x) = -\frac{\partial}{\partial x_n} \qquad \text{for } x \in \partial \Omega_k.$$

One way to pick such coordinate system is to take $C^\infty$ vector field which on $\partial \Omega_k$ coincide with inner unit normal vector to $\partial \Omega_k$. The flow generated by such vector field parameterizes some small collar neighborhood of $\partial \Omega_k$.

In this coordinate system we see that

$$\frac{\partial e_0}{\partial \nu_x}(x-y, y) = -\frac{\partial e_0}{\partial x_n}(x-y, y) = K_n \frac{\sum \delta_{jn}(y)(x_j - y_j)}{\left(\sum g_{jk}(y)(x_j - y_j)(x_k - y_k)\right)^{n/2}}.$$

Hence the kernel of the operator $K^*_{k,1}$ is equal to

$$(2.9) \qquad K^*_{k,1}(x,y) = K_n \psi(x,y) g(y)^{-1/2} \frac{x_n - y_n}{\left(\sum g_{jk}(y)(x_j - y_j)(x_k - y_k)\right)^{n/2}}.$$

This gives that for $x, y \in \partial\Omega_k$ $K^*_{k,1}(x,y) = 0$. Indeed, in this coordinates $x, y \in \partial\Omega$ can be written as $x = (x', 0)$ $y = (y', 0)$ for some $x', y' \in \mathbb{R}^{n-1}$, i.e., the numerator of (2.9) is zero.

If the metric tensor $g$ on $M$ is not smooth, we proceed similarly. We can find a sequence $g^\mu$ of *smooth* metric tensors on $M$ so that $g^\mu \to g$ uniformly (in $C^\gamma$ for all $\gamma < \alpha$) on $M$ as $\mu \to \infty$. Then for each $\mu$ we can pick a *smooth* coordinate system on a small neighborhood $U$ of any given point $z \in \partial\Omega$ by the process described above for the metric tensor $g^\mu$.

Since the metric tensor $g$ is close to $g^\mu$, we get from (2.8) that

$$(2.10) \quad \begin{aligned} g_{jn}(x) &= \delta_{jn} + h^\mu_{jn}(x), & \text{for } x \in \partial\Omega_k, \\ \nu(x) &= -A^\mu(x)\tfrac{\partial}{\partial x_n} + B^\mu(x)(\tfrac{\partial}{\partial x_1}, \tfrac{\partial}{\partial x_2}, \ldots, \tfrac{\partial}{\partial x_{n-1}})^T, & \text{for } x \in \partial\Omega_k. \end{aligned}$$

Here $A$ is scalar and $B$ a vector valued function. Also $h^\mu_{jn}(x) \to 0$, $A^\mu \to 1$ and $B^\mu \to 0$ uniformly (in $C^\gamma$ for all $\gamma < \alpha$) on $\partial\Omega_k$ as $\mu \to \infty$.

This means that for any given $\varepsilon > 0$ we can find $\mu$ big enough such that in the coordinate system corresponding to $g^\mu$ the kernel of the operator $K_{k,1}$ has small coefficients in the numerator. It follows by Proposition A.5 that the $\mathcal{L}(L^p)$ norm ($1 < p < \infty$) of the $K_{k,1}$ is small ($< \varepsilon$). Similar statement for Hardy spaces $\hbar^p$ follows from Proposition A.8. Finally, the same claim can be done also for the operators $K_{k,1}$. The argument is very similar.

Now we turn our attention to the operators $K_2$, $K_2^*$, $K_{k,2}$ and $K_{k,2}^*$. We treat again only one of them, namely $K_2$. The analysis for the operators $K_{k,2}$ is same, since for less singular kernels there is no difference between treating $\partial\Omega$ and $\partial\Omega_k$. Also the treatment of $K_2^*$ and $K_{k,2}^*$ is close to the treatment of $K_2$. The goal is to establish the following:

LEMMA 2.1. *Let $U$ be a small neighborhood of a point $x \in \partial\Omega$. Consider any smooth coordinates on $U$ and decompose the kernel $E(x,y)$ as in (1.35) in this coordinate system. Let $\operatorname{supp} \psi(x,y) \subset U \times U$. Then the operator*

$$(2.11) \qquad K_2 f(x) = \lim_{\varepsilon \to 0+} \int_{y \in O_\varepsilon} \psi(x,y) \frac{1}{\sqrt{g(x)}} \frac{\partial e_1}{\partial \nu_y}(y,x) f(y) \, d\sigma(y)$$

*is compact on $L^p(\partial\Omega)$ for any $1 < p < \infty$. Similar statement holds also for the operators $K_2^*$, $K_{k,2}$ and $K_{k,2}^*$.*

PROOF. Define operators $K_{2,\varepsilon}$ the same way we defined the operator $K_2$ in (2.11) but without taking the limit $\lim_{\varepsilon \to 0+}$. We claim that $K_{2,\varepsilon} \to K_2$ in the norm of $\mathcal{L}(L^p(\partial\Omega))$, $1 < p < \infty$ as $\varepsilon \to 0+$. To see this, let us estimate the difference

$(K_2 - K_{2,\varepsilon})f$ in the $L^1$ norm and then in $L^p$ norm for $p$ very big. We get:

$$\|(K_2 - K_{2,\varepsilon})f\|_{L^1(\partial\Omega)}$$

$$\leq \int_{\partial\Omega} \left| \int_{\partial\Omega} (1 - \chi_{O_\varepsilon}(x,y)) \frac{\psi(x,y)}{\sqrt{g(y)}} \frac{\partial e_1}{\partial \nu_y}(y,x) f(y) \, d\sigma(y) \right| d\sigma(x) \leq$$

(2.12)

$$\leq \int_{\partial\Omega} \int_{\partial\Omega} (1 - \chi_{O_\varepsilon}(x,y)) \frac{\psi(x,y)}{\sqrt{g(y)}} \left| \frac{\partial e_1}{\partial \nu_y}(y,x) \right| |f(y)| \, d\sigma(y) \, d\sigma(x) \leq$$

$$\leq C \int_{\partial\Omega} \left\| (1 - \chi_{O_\varepsilon}(.,y)) \frac{\partial e_1}{\partial \nu_y}(y,.) \right\|_{L^1(\partial\Omega)} \frac{1}{\sqrt{g(y)}} |f(y)| \, d\sigma(y) \leq C\lambda(\varepsilon) \|f\|_{L^1(\partial\Omega)},$$

$$\|(K_2 - K_{2,\varepsilon})f\|_{L^p(\partial\Omega)}^p$$

$$\leq \int_{\partial\Omega} \left| \int_{\partial\Omega} (1 - \chi_{O_\varepsilon}(x,y)) \frac{\psi(x,y)}{\sqrt{g(y)}} \frac{\partial e_1}{\partial \nu_y}(y,x) f(y) \, d\sigma(y) \right|^p d\sigma(x) \leq$$

(2.13)

$$\leq C \left( \int_{\partial\Omega} \left( \int_{\partial\Omega} (1 - \chi_{O_\varepsilon}(x,y)) \left| \frac{\partial e_1}{\partial \nu_y}(y,x) \right|^q d\sigma(y) \right)^{p/q} \|f\|_{L^p(\partial\Omega)}^p \, d\sigma(x) \right) \leq$$

$$\leq C \int_{\partial\Omega} \left\| (1 - \chi_{O_\varepsilon}(x,.)) \frac{\partial e_1}{\partial \nu}(.,x) \right\|_{L^q(\partial\Omega)}^{p/q} \|f\|_{L^p(\partial\Omega)}^p \, d\sigma(x) \leq C\lambda(\varepsilon)^{p/q} \|f\|_{L^p(\partial\Omega)}^p.$$

Here $q = p/(p-1)$, i.e., for $p$ big $q$ is close to 1. In (2.13) we used the Hölder inequality and then the estimate (1.23) on the $L^\infty$ norm of the gradient of $e_1(.,.)$ which gives that $\nabla_x e_1(x,y)$ belongs to $L^q$ for $q$ close to 1. Thus the functions $\lambda(.)$ in (2.12) and (2.13) tends to zero as $\varepsilon \to 0$. Interpolating between (2.12) and (2.13) yields that $K_{2,\varepsilon}f \to K_2 f$ in $L^p$ norm for any $1 \leq p < \infty$.

The second claim is that the operators $K_{2,\varepsilon}$ are compact for any $\varepsilon > 0$. From this compactness of $K_2$ follows. The operators $K_{2,\varepsilon}$ are compact, because any operator of the form

(2.14) $$Tf(x) = \int_{\partial\Omega} k(x,y) f(y) \, d\sigma(y)$$

with kernel $k(x,y)$ continuous on $\partial\Omega \times \partial\Omega$ is compact in $\mathcal{L}(L^p(\partial\Omega))$, $1 < p < \infty$.

In our case the kernel $k(x,y) = \frac{\psi(x,y)}{\sqrt{g(x)}} \frac{\partial e_1}{\partial \nu_y}(y,x) \chi_{O_\varepsilon}(x,y)$ is not continuous at the points $(x,y)$ where the geodesic distance $r(x,y) = \varepsilon$. But this is not a problem, since the characteristic function $\chi_{O_\varepsilon}$ of the set $O_\varepsilon$ can be approximated by a sequence of continuous functions

(2.15) $$\chi_k : \partial\Omega \times \partial\Omega \to \mathbb{R}, \quad \operatorname{supp} \chi_k \subset O_\varepsilon, \quad k = 1, 2, \ldots,$$

with the property that $\chi_k(x,.) \to \chi_{O_\varepsilon}(x,.)$ in any $L^q(\partial\Omega)$, $q < \infty$ uniformly in $x \in \partial\Omega$. Hence the operators with continuous kernels $\frac{\psi(x,y)}{\sqrt{g(x)}} \frac{\partial E}{\partial \nu_y}(x,y) \chi_k(x,y)$ converges in the operator norm to $K_{2,\varepsilon}$ as $k \to \infty$. $\square$

As we already indicated an immediate corollary of this lemma is:

COROLLARY 2.2. *Let $U$ be a small neighborhood of a point $x \in \partial\Omega$. Consider any smooth coordinates on $U$ and decompose the kernel $E(x,y)$ as in (1.35) in this coordinate system. Assume also that $\operatorname{supp} \psi(x,y) \subset U \times U$. Pick any $k \in \mathbb{N}$. Then the operators $K_{k,1}$, $K_{k,2}$ (and hence $K_k$), as well as, the operators $K_{k,1}^*$, $K_{k,2}^*$ (and hence $K_k^*$) are compact on $L^p(\partial\Omega_k)$ for any $1 < p < \infty$.*

PROOF. We have shown that for any given $\varepsilon > 0$ there is a special coordinate system on $U$ in which the operator norm of $K_{k,1}$ and $K_{k,1}^*$ is less than $\varepsilon$. By Lemma 2.2 in the same coordinate system the operators $K_{k,2}$ and $K_{k,2}^*$ are compact. These two things together guarantee that $K_k = K_{k,1} + K_{k,2}$ and $K_k^* = K_{k,1}^* + K_{k,2}^*$ are compact.

Now pick any smooth coordinate system. Lemma 2.1 still applies, that is the operators $\widetilde{K_{k,2}}$ and $\widetilde{K_{k,2}^*}$ obtained by the decomposition (1.35) in this coordinate system are again compact. Hence also $\widetilde{K_{k,1}} = K_k - \widetilde{K_{k,2}}$ and $\widetilde{K_{k,1}^*} = K_k^* - \widetilde{K_{k,2}^*}$ are compact, because they can be written as a difference of two compact operators. □

So far, we dealt mainly with $L^p$ spaces. The case of Hardy spaces $\hbar^p$ is similar but slightly more complicated. We begin with the following lemma.

LEMMA 2.3. *Let $(n-1)/n < p \leq 1$ and $r > (n-1)(p^{-1} - 1)$. Assume that the function $K : \partial\Omega \times \partial\Omega \to \mathbb{R}$ is continuous on $\partial\Omega \times \partial\Omega$ and Hölder continuous of modulus $r$ in the second variable, i.e., the function $y \mapsto K(x,y)$ is in $C^r$ uniformly for $x \in \partial\Omega$. Then the operator*

$$Tf(x) = \int_{\partial\Omega} K(x,y) f(y) \, d\sigma(y),$$

(2.16)
$$T : \hbar^p(\partial\Omega) \to \hbar^p(\partial\Omega)$$

*is well defined, bounded and compact.*

PROOF. We first show that for $f \in \hbar^p(\partial\Omega)$ we have $Tf \in L^\infty(\partial\Omega)$. Actually, this is a trivial observation. We just have to realize that the duality (1.29) and the fact that $r > \alpha = (n-1)(p^{-1}-1)$ imply

(2.17) $\quad |Tf(x)| \leq C \|K(x,.)\|_{C^\alpha} \|f\|_{\hbar^p(\partial\Omega)} \leq C \|K\|_{C^r} \|f\|_{\hbar^p(\partial\Omega)},$

for $(n-1)/n < p < 1$. If $p = 1$ we replace the $C^\alpha$ norm by bmo in (2.17).

(2.17) means that $Tf \in \hbar^p(\partial\Omega)$ boundedly, since $L^\infty(\partial\Omega) \subset \hbar^p(\partial\Omega)$. Now we look at the compactness of $T$. Take any $x, y, y', z \in \partial\Omega$ and examine the expression

(2.18) $\quad\quad\quad K(x,y') - K(x,y) - K(z,y') + K(z,y).$

Let us for simplicity denote the geodesic distance on $\partial\Omega$ between two points $x, y$ by $|x - y|$. Using the continuity of $K$ in the first variable we get

(2.19) $\quad\quad |K(x,y') - K(x,y) - K(z,y') + K(z,y)| \leq 2\omega(|x - z|).$

Here $\omega$ is the modulus of continuity of $K$. Similarly using the Hölder continuity of $K$ in the second variable we get

(2.20) $\quad\quad |K(x,y') - K(x,y) - K(z,y') + K(z,y)| \leq C|y' - y|^r.$

Combining (2.19) and (2.20) for any $\theta \in [0,1]$ we get

(2.21) $\quad |K(x,y') - K(x,y) - K(z,y') + K(z,y)| \leq C|y' - y|^{r\theta} \omega(|x - z|)^{1-\theta}.$

In particular choosing $\theta = \frac{\alpha}{r}$ we have for some $\varepsilon > 0$

$$(2.22) \qquad |K(x,y') - K(x,y) - K(z,y') + K(z,y)| \leq C|y'-y|^\alpha \omega(|x-z|)^\varepsilon.$$

This proves that the function $G(x,z,y) = K(x,y) - K(z,y)$ could be written as

$$(2.23) \qquad G(x,z,y) = \omega(|x-z|)^\varepsilon T(x,z,y),$$

where the function $T(x,z,y)$ is $C^\alpha$ Hölder continuous in the $y$ variable uniformly for all $x, z$. From this we get
$$(2.24)$$
$$|Tf(x) - Tf(z)| = \left| \int_{\partial\Omega} G(x,z,y) f(y) \, d\sigma(y) \right| \leq C\omega(|x-z|)^\varepsilon \|T(x,z,.)\|_{C^\alpha} \|f\|_{\hbar^p(\partial\Omega)}.$$

If $p = 1$ we replace the norm $C^\alpha$ by bmo. Since our sequence $(f_n)$ is bounded in $\hbar^p(\partial\Omega)$ we get by (2.17) and (2.24) that

$$(2.25) \qquad \|Tf_n\|_{L^\infty(\partial\Omega)} \leq C \quad \text{and} \quad |Tf_n(x) - Tf_n(z)| \leq C\omega(|x-z|)^\varepsilon.$$

This means $(Tf_n)$ is sequence of uniformly bounded and equicontinuous function. By the Ascoli theorem it has a subsequence which converges in $C(\partial\Omega)$. Since $C(\partial\Omega) \subset \hbar^p(\partial\Omega)$ our subsequence is convergent also there. This concludes the proof. $\qquad\square$

Now we can prove a lemma analogous to Lemma 2.1.

LEMMA 2.4. *Assume that $\partial\Omega \in C^1$. Let $U$ be a small neighborhood of a point $x \in \partial\Omega$. Consider any smooth coordinates on $U$ in which we can write $\Omega \cap U$ as in (2.3) and decompose the kernel $E(x,y)$ as in (1.35) in this coordinate system. Let $\mathrm{supp}\, \psi(x,y) \subset U \times U$. Then the operator*

$$(2.26) \qquad K_2^* f(x) = \lim_{\varepsilon \to 0+} \int_{y \in O_\varepsilon} \psi(x,y) \frac{1}{\sqrt{g(y)}} \frac{\partial e_1}{\partial \nu_x}(x,y) f(y) \, d\sigma(y)$$

*is compact on $\hbar^p(\partial\Omega)$ for any $\max\{\frac{n-1}{n}, \frac{n-1}{\alpha+n-1}\} < p \leq 1$. Similar statement holds also for the operators $K_{k,2}^*$. If we also assume that the boundary $\partial\Omega$ is of class $C^{1+\alpha}$ we have the same statement for the operators $K_2, K_{k,2}$.*

PROOF. The assumption $\partial\Omega \in C^1$ implies that the part of the boundary $\partial\Omega \cap U$ can be written as $x = (x', \varphi(x'))$ where $x' \in \mathbb{R}^{n-1}$ and $\varphi$ is a $C^1$ function. Hence, there is a number $m$ such that $|\nabla \varphi| \leq m$. By $e_n$ we denote the vector $(0,0,\ldots,0,1) \in \mathbb{R}^n$. Observe that if we take $\widetilde{x} = x + te_n$, then

$$(2.27) \qquad r(x, \widetilde{x}) \approx r(\widetilde{x}, \partial\Omega) \approx t, \qquad \text{for any } x \in \partial\Omega \text{ and } t \text{ small}.$$

Here $r(.,.)$ again means the geodesic distance on $M$.

For $\varepsilon \neq 0$ we define the operator $K_{2,\varepsilon}^*$ from $\hbar^p(\partial\Omega)$ to $\hbar^p(\partial\Omega)$ by

$$(2.28) \qquad K_{2,\varepsilon}^* f(x) = \int_{\partial\Omega} \psi(x,y) \frac{1}{\sqrt{g(y)}} \frac{\partial e_1}{\partial \nu_x}(x + \varepsilon e_n, y) f(y) \, d\sigma(y).$$

We claim that for $\varepsilon \neq 0$ Lemma 2.3 applies and hence $K_{2,\varepsilon}^*$ is a family of compact operators from $\hbar^p(\partial\Omega)$ to $\hbar^p(\partial\Omega)$, provided $\max\{\frac{n-1}{n}, \frac{n-1}{\alpha+n-1}\} < p \leq 1$. Verifying this is not difficult. By Proposition 2.4 of [24] the kernel $E(x,y)$ is $C_{\mathrm{loc}}^{1+\alpha}$ of the diagonal. Hence (1.21), (1.22) and (1.35) give us that the kernel of $K_{2,\varepsilon}^*$ is continuous and of class $C^\alpha$ in the second variable provided $\varepsilon \neq 0$, i.e., we stay away from the singularity.

The claim of Lemma 2.4 will be established if we prove that $K_2^*$ is a limit of $K_{2,\varepsilon}^*$ in the norm of $\mathcal{L}(\hbar^p(\partial\Omega))$ as $\varepsilon \to 0$.

For this we need a little more regularity on $e_1(x,y)$ than follows from (1.23). We get it by rereading the proof of Lemma 2.5 in [24]. Namely, let $|x_0 - y| = 2\rho$, we want to estimate $e_1(x,y)$ on $\{x : |x - x_0| \leq \rho\}$. We shift coordinates so $x_0 = 0$ and introduce dilation operators

$$(2.29) \qquad u_\rho(x) = u(\rho x), \qquad |x| \leq 1.$$

If $u(x) = e_1(x,y)$ for $|x| \leq \rho$, then (2.76)-(2.80) of [24] yields

$$(2.30) \quad \|u_\rho\|_{H^{s,q}(B_{1/2})} \leq C(s,q,\delta)\rho^{-(n-2-\alpha+\delta)}, \qquad \forall s < 1+\alpha, \ q < \infty, \ \delta > 0.$$

Hence for any $\delta > 0$

$$(2.31) \qquad \|\nabla_x u_\rho\|_{C^{\alpha-\delta}(B_{1/2})} \leq C_\delta \rho^{-(n-2-\alpha+\delta)}.$$

We actually do not need Hölder regularity of the order of $\alpha - \delta$. Therefore by possibly making $\delta > 0$ smaller so that $\alpha - \delta > \delta$ we can get

$$(2.32) \qquad \|\nabla_x u_\rho\|_{C^\delta(B_{1/2})} \leq C_\delta \rho^{-(n-2-\alpha+\delta)}.$$

This means that for $|x - x_0| \leq \frac{1}{4}|x_0 - y|$ we get

$$(2.33) \qquad |\nabla_x e_1(x,y) - \nabla_x e_1(x_0,y)| \leq C_\delta |x_0 - y|^{-(n-1-\alpha+2\delta)} |x - x_0|^\delta.$$

In our setting we want to take $x_0, y \in \partial\Omega$ and $x = x_0 + \varepsilon\nu(x_0)$. So, by (1.23) even for $|x - x_0| > \frac{1}{4}|x_0 - y|$ (2.33) remains true, because then $|x - x_0| \approx |x - y|$. Indeed, in this case

$$|\nabla_x e_1(x,y) - \nabla_x e_1(x_0,y)| \leq C(|x-y|^{-(n-1-\alpha+\delta)} + |x_0-y|^{-(n-1-\alpha+\delta)}) \leq$$
$$\leq C|x_0-y|^{-(n-1-\alpha+\delta)} \leq C|x_0-y|^{-(n-1-\alpha+2\delta)}|x_0-y|^\delta \leq$$
$$(2.34)$$
$$\leq C|x_0-y|^{-(n-1-\alpha+2\delta)}|x-x_0|^\delta.$$

Finally, (2.33) and (2.34) allow us to estimate the difference between the kernels $K_2$ and $K_{2,\varepsilon}$. We get

$$|T_\varepsilon(x,y)| = |K_{2,\varepsilon}^*(x,y) - K_2^*(x,y)| =$$
$$= \psi(x,y)\frac{1}{\sqrt{g(y)}}\left|\frac{\partial e_1}{\partial \nu_x}(x+\varepsilon e_n, y) - \frac{\partial e_1}{\partial \nu_x}(x,y)\right| \leq$$
$$(2.35) \qquad \leq C_\delta \varepsilon^\delta |x-y|^{-(n-1-\alpha+2\delta)}.$$

This is the estimate we sought. As we have observed before an operator with a kernel $T(x,y)$ which has singularity of the order $|y-x|^{-(n-1-\alpha+2\delta)}$ on diagonal maps $\hbar^p(\partial\Omega)$ into itself boundedly. Notice also that off the singularity the operator $T_\varepsilon$ is continuous in the first variable and of class $C^\alpha$ in the second one. As we let $\varepsilon \to 0+$ the constant $C_\delta \varepsilon^\delta$ goes to zero and hence the family $(T_\varepsilon)_{\varepsilon > 0}$ goes to to zero in the operator norm. This establishes the lemma for $K_2^*$. The proof for the operator $K_2$ goes exactly as the proof above. The additional assumption that

$\partial\Omega$ belongs to $C^{1+\alpha}$ is required in this case, because $\hbar^p(\partial\Omega)$ is not a module over $C(\partial\Omega)$, but it is over $C^\alpha(\partial\Omega)$. □

As a corollary we get:

COROLLARY 2.5. *Let $\partial\Omega \in C^1$ and $U$ be a small neighborhood of a point $x \in \partial\Omega$. Consider any smooth coordinates on $U$ in which we can write $\Omega_k \cap U$ as in (2.3) and decompose the kernel $E(x,y)$ as in (1.35) in this coordinate system. Assume also that supp $\psi(x,y) \subset U \times U$. Pick any $k \geq 1$. Then the operators $K_2^*$, $K_{k,1}^*$, $K_{k,2}^*$ (and hence $K_k^*$) are compact for any $\max\{\frac{n-1}{n}, \frac{n-1}{\alpha+n-1}\} < p \leq 1$.*

*If we also assume $\partial\Omega \in C^{1+\alpha}$ then the operators $K_2$, $K_{k,1}$, $K_{k,2}$ (and hence $K_k$) are compact on $\hbar^p(\partial\Omega)$ for the same range of $p$.*

The final point that will establish compactness of the operators $K$, $K^*$ is the convergence $K_k \to K$ and $K_k^* \to K^*$, as $k \to \infty$. First, we again consider the technically easier $L^p$ case.

PROPOSITION 2.6. *Assume that the metric tensor on $M$ is of class $C^\alpha$. Let $1 < p < \infty$ and $\partial\Omega \in C^1$. Then the operators*

(2.36) $$K, K^* : L^p(\partial\Omega) \to L^p(\partial\Omega)$$

*are well defined, bounded and compact.*

PROOF. We do the proof for the operator $K$ only. By Lemma 2.2 it suffices to show compactness of $K_1$. To simplify the notation we put $\varphi_0 := \varphi$ and $K_{0,1} := K_1$.

Recall, that we have localized the problem to some neighborhood $U$ where (2.3) and (2.4) hold. By Corollary 2.2, the operators $K_{k,1}$ are compact on $L^p$ for any $k = 1, 2, \ldots$ and $1 < p < \infty$. The another way to look at these operators is to think of them as operators on $L^p(\mathbb{R}^{n-1})$. This can be done by putting

(2.37)
$$\overline{K}_{k,1}f(x') = \lim_{\varepsilon \to 0+} \int_{|x'-y'|>\varepsilon} \psi((x',\varphi_k(x')),(y',\varphi_k(y'))) \frac{1}{\sqrt{g(x',\varphi_k(x'))}} \times$$
$$\left(\sum_{i=1}^n \frac{\partial e_0}{\partial z_i}((x'-y',\varphi_k(x')-\varphi_k(y')),(x',\varphi_k(x')))\nu_k^i(y')\right) f(y')\sqrt{\overline{g}(y',\varphi_k(y'))}\,dy',$$

for $k = 0, 1, 2, \ldots$. We have that (2.37) is a compact operator on $L^p(\mathbb{R}^{n-1})$ if and only if $K_{k,1}$ is compact on $L^p(\partial\Omega_k)$. Here $\nu_k^i(y')$ is the $i$-th component of the outer normal to the curve $\partial\Omega_k$ at the point $(y',\varphi_k(y'))$, $\frac{\partial e_0}{\partial z_i}$ is a partial derivative of $e_0(z,y)$ with respect to the first variable and $\overline{g}$ is the metric tensor on $\partial\Omega_k$ inherited from $M$. Also clearly, because $\nu_k^i$ depends continuously on the metric tensor $g_{ij}$ and on $\nabla\varphi_k$, as $k \to \infty$, $\nu_k^i \to \nu_0^i$ uniformly in $y'$, where $\nu_0^i(y')$ is the $i$-th component of the outer normal to the curve $\partial\Omega$ at the point $(y',\varphi_0(y'))$. The same could be said about $g(y',\varphi_k(y'))$, i.e., $g(y',\varphi_k(y')) \to g(y',\varphi_0(y'))$ uniformly in $y'$ because of (2.4).

The advantage of looking at $\overline{K}_{k,1}$ instead of $K_{k,1}$ is that $\overline{K}_{k,1}$ act on the same space for each $k$. Now, Proposition A.4. of Appendix A which is a modification of a result of A. P. Calderon in [**2**] and Theorem 4 of [**3**] used on (2.37) gives us the convergence of $\overline{K}_{k,1}$ to $\overline{K}_1$ in the $\mathcal{L}(L^p(\mathbb{R}^{n-1}))$ norm for $1 < p < \infty$.

Subsequently, the operator $\overline{K}_{0,1}$ is compact on $L^p(\mathbb{R}^{n-1})$ and therefore the operator $K_1 = K_{0,1}$ is compact on $L^p(\partial\Omega)$. □

Now, a similar statement for Hardy spaces.

PROPOSITION 2.7. *Assume that the metric tensor on $M$ is of class $C^\alpha$. Let $1 < p < \infty$ and $\partial\Omega \in C^1$. Let $\max\{\frac{n-1}{n}, \frac{n-1}{\alpha+n-1}\} < p \leq 1$ and $\partial\Omega \in C^1$. Then the operator*

$$(2.38) \qquad K^* : \hbar^p(\partial\Omega) \to \hbar^p(\partial\Omega)$$

*is well defined, bounded and compact. Moreover if the boundary $\partial\Omega$ is of class $C^{1+\alpha}$ same is true about the operator*

$$(2.39) \qquad K : \hbar^p(\partial\Omega) \to \hbar^p(\partial\Omega).$$

PROOF. We again do our work for only one of these operators, namely $K^*$. First, we pick $\Theta$ to be a smooth vector field on $M$ transverse to $\partial\Omega$ pointing inside $\Omega$. Flow $\mathcal{F}_t$ generated by such vector field allows us to identify $\partial\Omega$ with $\partial\Omega_k$ (at least for $k$ big), since for such $k$ the vector field $\Theta$ is also transversal to $\partial\Omega_k$ and therefore for each point $x \in \partial\Omega$ there is a exactly one point $\mathcal{F}_t x \in \partial\Omega_k$ for a certain $t$ and vice versa. Via such identification, for any $f \in \hbar^p(\partial\Omega)$ the function $f\rho_k$ can be thought of as a function on $\partial\Omega_k$ and moreover $f\rho_k \in \hbar^p(\partial\Omega_k)$. Here $\rho_k = \frac{d\sigma}{d\sigma_k}$ is the Radon-Nikodym derivative of surface measures we consider on $\partial\Omega$ and $\partial\Omega_k$.

Fix $f \in \hbar^p(\partial\Omega)$ with norm bounded by 1 and consider $u = \mathcal{S}f$ for $f \in \hbar^p(\partial\Omega)$ and $x \in \Omega$. For simplicity consider just the case when $V = 0$ in $\Omega$. The other case $V \neq 0$ is similar. By (1.15) and (1.34) we have that

$$(2.40) \qquad f_0 = \partial_\nu \mathcal{S}f = (-\tfrac{1}{2}I + K^*)f \in \hbar^p(\partial\Omega).$$

We want to apply Proposition A.8. In order to do that, we again localize the operator $K^*$ onto a neighborhood $U$ by a partition of unity. Then Proposition A.8 applies to the main piece $\nabla e_0(x-y, y)$ in the decomposition (1.35) of the kernel of $K^*$, while the contribution of the remaining term $\nabla e_1(x, y)$ dealt with using (1.23). This yields

$$(2.41) \qquad \|(\nabla(\mathcal{S}_k(f\rho_k) - \mathcal{S}f))^*\|_{L^p(\partial\Omega)} \leq \varepsilon_k, \qquad \text{as } k \to \infty.$$

Here $\mathcal{S}_k$ is the corresponding single layer potential on $\Omega_k$ and $\varepsilon_k \searrow 0$ as $k \to \infty$ does not independent on $f$, as long as, $f$ is bounded in the norm by one.

The estimate (2.41) should be understood in the following sense. The gradient $\nabla(\mathcal{S}_k(f\rho_k))$ is defined on $\Omega_k$, whereas $\nabla(\mathcal{S}f)$ is defined on $\Omega$. Hence, if we want to compare their maximal functions, we have to do is as in Proposition A.8. For any point $x_0 \in \Omega$ there is a unique $x_0' \in \Omega_k$ for which

$$(2.42) \qquad x_0' = \mathcal{F}_t x_0 \qquad \text{for some } t > 0,$$

where the flow $\mathcal{F}_t$ is as above. So, if $\gamma(x_0)$ is a nontangetial approach region to $\Omega$ at $x_0$, then $\gamma'(x_0') = \mathcal{F}_t(\gamma(x_0))$ is a nontangential approach region to $\Omega_k$ at $x_0'$. Hence, if we take $\nabla(\mathcal{S}_k(f\rho_k))$ at $\gamma'(x_0')$ where it is well defined, pull it back to $\gamma(x_0)$ via the flow, we finally get two objects $\nabla(\mathcal{S}_k(f\rho_k))$ and $\nabla(\mathcal{S}f)$ well defined on $\gamma(x_0)$. Then we can compute (2.41).

In particular, we take the normal derivative $\frac{\partial}{\partial \nu}$ we get from (2.41)

$$\|(\tfrac{\partial}{\partial \nu}(\mathcal{S}_k(f\rho_k) - \mathcal{S}f))^*\|_{L^p(\partial\Omega)} \leq \varepsilon_k, \qquad \text{as } k \to \infty. \tag{2.43}$$

Again the meaning of (2.43) should be clarified. With a point $x_0 \in \partial\Omega$ fixed $\frac{\partial}{\partial \nu}(x_0)$ is a well defined vector at $x_0$ which could be simply extended by a parallel transport to a neighborhood $U$ of $x_0$. This means

$$\tfrac{\partial}{\partial \nu}(\mathcal{S}_k(f\rho_k) - \mathcal{S}f) \tag{2.44}$$

is well defined on $\Omega \cap U$ and thus we can evaluate the maximal operator in (2.43).

If $\frac{\partial}{\partial \nu_k}$ denotes the outer normal to $\partial\Omega_k$ from the assumption about our domains $\Omega_k$ we get that it can be written as

$$\frac{\partial}{\partial \nu_k} = A_k \frac{\partial}{\partial \nu} + B_k \nabla_{T_k}, \tag{2.45}$$

where $\nabla_{T_k}$ denotes a tangential derivative with respect to $\partial\Omega_k$, $A_k$ is a real and $B_k$ a vector valued function and $A_k \to 1$ and $B_k \to 0$ in the $L^\infty$ norm as $k \to \infty$.

Combining (2.45) with (2.43) and the fact that from (2.41) we get a uniform $L^p$ bound on $\nabla_{T_k} \mathcal{S}_k(f\rho_k)$ we conclude

$$\|(\tfrac{\partial}{\partial \nu_k}(\mathcal{S}_k(f\rho_k)) - \tfrac{\partial}{\partial \nu}(\mathcal{S}f))^*\|_{L^p(\partial\Omega)} \leq \varepsilon'_k, \qquad \text{as } k \to \infty \tag{2.46}$$

for a sequence $(\varepsilon'_k)_{n \in N}$ converging to zero. (2.46) is again understood in the sense explained above, i.e., via the flow $\mathcal{F}_t$. In particular, we have

$$(\tfrac{\partial}{\partial \nu_k}(\mathcal{S}_k(f\rho_k)))^* \in L^p, \tag{2.47}$$

and hence due to result of Wilson [**31**] it follows that

$$\tfrac{\partial}{\partial \nu_k}\mathcal{S}_k(f\rho_k) \in \hbar^p(\partial\Omega_k), \tag{2.48}$$

or

$$\tfrac{1}{\rho_k}\tfrac{\partial}{\partial \nu_k}\mathcal{S}_k(f\rho_k) \in \hbar^p(\partial\Omega). \tag{2.49}$$

This means that (2.49) and (2.47) are equivalent. Hence $(\tfrac{\partial}{\partial \nu_k}(\mathcal{S}_k(f\rho_k)) - \tfrac{\partial}{\partial \nu}(\mathcal{S}f))^*$ could be seen as a function on $\partial\Omega$ in $L^p$ and therefore again due to [**31**] this is equivalent to

$$\|\tfrac{1}{\rho_k}\tfrac{\partial}{\partial \nu_k}(\mathcal{S}_k(f\rho_k)) - \tfrac{\partial}{\partial \nu}(\mathcal{S}f)\|_{\hbar^p(\partial\Omega)} \approx \|(\tfrac{\partial}{\partial \nu_k}(\mathcal{S}_k(f\rho_k)) - \tfrac{\partial}{\partial \nu}(\mathcal{S}f))^*\|_{L^p(\partial\Omega)} \leq \varepsilon'_k \to 0. \tag{2.50}$$

Notice that

$$\tfrac{1}{\rho_k}K_k^*(f\rho_k) - K^*f = \tfrac{1}{\rho_k}(-\tfrac{1}{2}I + K_k^*)(f\rho_k) - (-\tfrac{1}{2}I + K^*)f. \tag{2.51}$$

Combining (2.50) and (2.51) finally yields

$$\|\tfrac{1}{\rho_k}K_k^*(f\rho_k) - K^*f\|_{\hbar^p(\partial\Omega)} = \|\tfrac{1}{\rho_k}\tfrac{\partial}{\partial \nu_k}(\mathcal{S}_k(f\rho_k)) - \tfrac{\partial}{\partial \nu}(\mathcal{S}f)\|_{\hbar^p(\partial\Omega)} \leq \varepsilon'_k \to 0. \tag{2.52}$$

This is the desired result. Now the compactness of $K_k^*$ gives us compactness of $K^*$. $\square$

## 2.2 Invertibility of $\pm\frac{1}{2}I + K$, $\pm\frac{1}{2}I + K^*$

In the second part of this chapter we apply Propositions 2.6 and 2.7.

PROPOSITION 2.8. *Let $\partial\Omega \in C^1$. The operators*

(2.53) $\quad\begin{aligned}\pm\tfrac{1}{2}I + K, \pm\tfrac{1}{2}I + K^* &: L^p(\partial\Omega) \to L^p(\partial\Omega) & 1 < p < \infty, \\ \pm\tfrac{1}{2}I + K^* &: \hbar^p(\partial\Omega) \to \hbar^p(\partial\Omega) & \max\{\tfrac{n-1}{n}, \tfrac{n-1}{\alpha+n-1}\} < p \leq 1\end{aligned}$

*are Fredholm of index zero. If $\partial\Omega \in C^{1+\alpha}$ then also*

(2.54) $\quad\pm\tfrac{1}{2}I + K : \hbar^p(\partial\Omega) \to \hbar^p(\partial\Omega)$

*is Fredholm of index zero for the same range of p.*

PROOF. Since $\pm\frac{1}{2}I$ is Fredholm and $K$, $K^*$ are compact the Fredholmness of the operators (2.53) and (2.54) is guaranteed. The claim that these operators have index zero then follows from the fact that any compact perturbation of identity has index zero. □

THEOREM 2.9. *Assume the same as in Proposition 2.8. The maps*

(2.55) $\quad\begin{aligned}\tfrac{1}{2}I + K, \tfrac{1}{2}I + K^* &: L^p(\partial\Omega) \to L^p(\partial\Omega) & 1 < p < \infty, \\ \tfrac{1}{2}I + K, \tfrac{1}{2}I + K^* &: \hbar^p(\partial\Omega) \to \hbar^p(\partial\Omega) & \max\{\tfrac{n-1}{n}, \tfrac{n-1}{\alpha+n-1}\} < p \leq 1\end{aligned}$

*are isomorphisms. If $V > 0$ on a set of positive measure in $\Omega$ then $-\frac{1}{2}I + K$, $-\frac{1}{2}I + K^*$ are also invertible in this range of p. If $V = 0$ on $\overline{\Omega}$ then*

(2.56) $\quad\begin{aligned}-\tfrac{1}{2}I + K^* &: L_0^p(\partial\Omega) \to L_0^p(\partial\Omega) & 1 < p < \infty, \\ -\tfrac{1}{2}I + K^* &: \hbar_{at}^p(\partial\Omega) \to \hbar_{at}^p(\partial\Omega) & \max\{\tfrac{n-1}{n}, \tfrac{n-1}{\alpha+n-1}\} < p \leq 1\end{aligned}$

*are isomorphisms where $L_0^p(\partial\Omega)$ consists of $L^p$ functions integrating to zero.*

PROOF. The special case of (2.55) when $p = 2$ has been established in Proposition 4.1 of [**22**] for the operator $\frac{1}{2}I + K^*$. Duality gives the same result for the other operator. Now let $\max\{\frac{n-1}{n}, \frac{n-1}{\alpha+n-1}\} < p \leq 1$, we want to show invertibility of (2.55) in $\hbar^p(\partial\Omega)$. It suffices to show that the operators $\frac{1}{2}I + K^*$, $\frac{1}{2}I + K$ have dense range in $\hbar^p(\partial\Omega)$. But this is a consequence of the fact that the invertibility of this operator for $p = 2$ implies that the range of $\frac{1}{2}I + K^*$, $\frac{1}{2}I + K$ contains all $L^2(\partial\Omega)$ functions. The space $L^2(\partial\Omega)$ is dense in $\hbar^p(\partial\Omega)$.

Now for any $p > 1$ since $L^p \subset \hbar^1$ and on $\hbar^1(\partial\Omega)$ these operators are invertible (hence their kernel is zero), it follows that operators (2.55) have zero kernel in $L^p$ and therefore are invertible.

Now, let $V > 0$ on a set of positive measure in $\Omega$. Again the case of invertibility of operators $-\frac{1}{2}I + K$, $-\frac{1}{2}I + K^*$ for $L^2(\partial\Omega)$ has been established in Proposition 4.6 of [**22**]. By the same argument as above we are again able to prove that these operators have dense range in any $\hbar^p$, $\max\{\frac{n-1}{n}, \frac{n-1}{\alpha+n-1}\} < p \leq 1$. Then the fact that $L^p \subset \hbar^1$ for $p > 1$ and injectivity on $\hbar^1$ establishes the invertibility on $L^p$.

Finally, let $V = 0$ on $\overline{\Omega}$. Again the $L^2$ case is dealt with in Proposition 4.6 of [**22**]. It follows, that in this case both the kernel and cokernel of $-\frac{1}{2}I + K^*$ in $L^2$ contain just the constants. By a density argument then the kernel and cokernel of $-\frac{1}{2}I + K^*$ contain the constants in any $L^p$ and $\hbar^p$ space we consider. Take again any

$\max\{\frac{n-1}{n}, \frac{n-1}{\alpha+n-1}\} < p \le 1$. If we show that the cokernel of this operator in $\hbar^p(\partial\Omega)$ is exactly one dimensional we will be done. The argument goes exactly as above; the range of $-\frac{1}{2}I + K^*$ contains $L_0^2(\partial\Omega)$ and therefore is dense in $\hbar^p_{\text{at}}(\partial\Omega)$. From this we also get that the kernel of $-\frac{1}{2}I + K^*$ contains only constants in $\hbar^1(\partial\Omega)$, and hence also in $L^p(\partial\Omega)$ for any $p > 1$. □

From Theorem 2.9 by duality we also get:

THEOREM 2.10. *Assume that $\partial\Omega \in C^1$. For any $0 < r < \min\{1, \alpha\}$ the maps*

(2.57) $$\tfrac{1}{2}I + K : C^r(\partial\Omega) \to C^r(\partial\Omega)$$

*and*

(2.58) $$\tfrac{1}{2}I + K : bmo(\partial\Omega) \to bmo(\partial\Omega)$$

*are isomorphisms. If $\partial\Omega \in C^{1+\alpha}$ the same result is true about*

(2.59) $$\tfrac{1}{2}I + K^* : C^r(\partial\Omega) \to C^r(\partial\Omega)$$

*and*

(2.60) $$\tfrac{1}{2}I + K^* : bmo(\partial\Omega) \to bmo(\partial\Omega).$$

*If $V > 0$ on a set of positive measure in $\Omega$ then $-\frac{1}{2}I + K^*$ is invertible on $bmo(\partial\Omega)$ and $C^r(\partial\Omega)$, granted $0 < r < \min\{1, \alpha\}$ and $\partial\Omega \in C^{1+\alpha}$. If $V = 0$ on $\overline{\Omega}$ then*

(2.61) $$-\tfrac{1}{2}I + K^* : C^r(\partial\Omega)\big/(1) \to C^r(\partial\Omega)\big/(1)$$

*and*

(2.62) $$-\tfrac{1}{2}I + K^* : bmo(\partial\Omega)\big/(1) \to bmo(\partial\Omega)\big/(1)$$

*are isomorphisms. Here $C^r(\partial\Omega)\big/(1)$ means a Banach space $C^r(\partial\Omega)$ modulo constants.*

PROOF. The only remaining part is to establish (2.61) and (2.62) since these results does not follow immediately from duality. Nevertheless duality gives us that $-\frac{1}{2}I + K^*$ on $C^r$ and bmo is Fredholm of index zero and the previous result for $L^p$ gives that its kernel and cokernel contains only constants. □

CHAPTER 3

# The Dirichlet Problem

## 3.1. $L^p$ boundary data

We retain all hypothesis on $M$, $\Omega$ and $L = \Delta - V$ we made before. Let us explicitly assume that the metric tensor is of class $C^\alpha$ for some $0 < \alpha < 1$.

THEOREM 3.1. *Let $\partial\Omega \in C^1$. Given $f \in L^p(\partial\Omega)$, $1 < p \leq \infty$ there exists a unique function $u \in C^{1+\alpha}_{\text{loc}}(\Omega)$ satisfying*

(3.1) $\qquad Lu = 0 \text{ in } \Omega, \qquad u^* \in L^p(\partial\Omega), \qquad u\big|_{\partial\Omega} = f \in L^p(\partial\Omega),$

*the limit on $\partial\Omega$ taken in the nontangential a.e. sense. Moreover, $u$ is representable in the form*

(3.2) $\qquad\qquad u = \mathcal{D}((\tfrac{1}{2}I + K)^{-1}f) \text{ in } \Omega,$

*and there is a uniform estimate*

(3.3) $\qquad\qquad \|u^*\|_{L^p(\partial\Omega)} \leq C_p \|f\|_{L^p(\partial\Omega)}.$

PROOF. By Theorem 2.9 $\tfrac{1}{2}I + K$ is invertible and therefore the function $u$ defined by (3.2) due to (1.13) and (1.18) solves (3.1) and satisfies (3.3) for any $1 < p < \infty$. If $p = \infty$ the same argument as used in Proposition 5.7 of [**22**] shows that the $L^2$ solution of (3.1) with $f \in L^\infty(\partial\Omega)$ satisfies

$$\|u\|_{L^\infty(\Omega)} \leq \|f\|_{L^\infty(\partial\Omega)}.$$

It remains to establish uniqueness. For $p > 2 - \varepsilon$ this is done in Proposition 9.1 of [**23**]. Before we prove uniqueness for our range of $p$ we first need to do a little more work.

Let $\Theta$ be a smooth vector field on $M$ which is transversal to $\partial\Omega$ and points into $\Omega$. Denote by $\mathcal{F}_t$ the flow generated by $\Theta$ on $M$ and introduce domains $\Omega_t$ for $t > 0$ small by mapping $\Omega$ onto $\Omega_t$ via the flow $\mathcal{F}_t$. Clearly $\Omega$ and $\Omega_t$ are diffeomorphic and $\Omega_t$, $t \searrow 0$ are increasing domains $\overline{\Omega_t} \subset\subset \Omega$ approximating $\Omega$ so that $\nu_t$ the outer unit normal to $\partial\Omega_t$ converges in $C^1$ to $\nu$ the outer unit normal to $\partial\Omega$ as $t \searrow 0$. Now let $g_t = \mathcal{F}_t^* g$ denote the metric tensor on $M$ that is the pull-back of the original metric $g$ under $\mathcal{F}_t$. Also let $\Delta_t$ denote the Laplace operator on $M$ for the metric $g_t$ and $d\sigma_t$ the surface measure on $\partial\Omega$ induced by this metric tensor. Set $v_t(x) = \mathcal{F}_t^* v(x) = v(\mathcal{F}_t x)$ and $L_t = \Delta_t + V_t$.

Let $u$ be any function solving (3.1) with boundary data $u\big|_{\partial\Omega} = 0$. If we show that $u = 0$ on $\Omega$, the uniqueness in Theorem 3.1 will follow. By interior regularity $u \in C^{1+\alpha}_{\text{loc}}(\Omega)$ and therefore $u\big|_{\partial\Omega_t}$ is a continuous function for all $t > 0$. Hence, also $u_t\big|_{\partial\Omega}$ is continuous. It has been established in [**22**] that for such functions the solution given by (3.2) coincides with the classical solution (Poisson integral) to the problem:

(3.4) $\qquad Lu = 0 \text{ in } \Omega, \qquad u \in C(\overline{\Omega}), \qquad u|_{\partial\Omega} = f \in C(\partial\Omega),$

which is unique. Hence we get that

(3.5) $\qquad u_t(x) = u(\mathcal{F}_t x) = \mathcal{D}_t((\tfrac{1}{2}I + K_t)^{-1} u_t|_{\partial\Omega}), \qquad \text{for } x \in \Omega \text{ and } t > 0,$

where $\mathcal{D}_t$ and $K_t$ are defined as in the chapter 1, with the metric tensor $g_t$. We want to establish the following lemma.

LEMMA 3.2. *For any $1 < p < \infty$ the operators $K_t$ converge to $K$ and the operators $\mathcal{D}_t$ converge to $\mathcal{D}$ in the $L^p$ norm as $t \searrow 0$. More precisely*

(3.6)
$$\|K - K_t\|_{\mathcal{L}(L^p(\partial\Omega), L^p(\partial\Omega))} \to 0 \qquad \text{as } t \searrow 0,$$
*and for any $x \in \Omega$:* $\qquad \|\mathcal{D}(x) - \mathcal{D}_t(x)\|_{\mathcal{L}(L^p(\partial\Omega), \mathbb{R})} \to 0 \qquad \text{as } t \searrow 0.$

*Moreover, the same is true in the Hardy space norm $\hbar^p$ for $\max\{\frac{n-1}{n}, \frac{n-1}{\alpha+n-1}\} < p \leq 1$, provided $\partial\Omega \in C^{1+\alpha}$.*

PROOF. The results from Chapter 2, namely an equivalent of (2.52) for the operator $K$ gives us this result for Hardy spaces. Consider therefore just the $L^p$ case. From the proof of Proposition 2.6 it follows that in the decomposition (2.7) of the operator $K = K_1 + K_2$ we can concentrate on the part $K_2$ only, since the convergence (3.6) of the part $K_1$ has been established there.

We have to clarify what norm on $L^p(\partial\Omega)$ we have in mind, since $\partial\Omega$ is now equipped with measures $d\sigma$ and $d\sigma_t$. However with the assumption of having a $C^1$ boundary $\partial\Omega$ and a $C^\alpha$ metric tensor, the Radon-Nikodym derivative $\rho_t = d\sigma_t/d\sigma$ of $d\sigma_t$ with respect to the original surface measure $d\sigma$ is continuous and tends to 1 in $L^\infty$ as $t \searrow 0$. (For the Hardy spaces the assumption $\partial\Omega \in C^{1+\alpha}$ gives us $\rho_t \in C^\alpha$, i.e., we can freely multiply and divide by $\rho_t$, since $\hbar^p(\partial\Omega)$ for our range of $p$ is a module over $C^\alpha$).

This implies that all norms on $L^p(\partial\Omega)$ and $\hbar^p(\partial\Omega)$ with respect to the measures $d\sigma$ and $d\sigma_t$ are uniformly equivalent. We can write the operator $K_{2,t}$ on $\partial\Omega$ as

(3.7) $\qquad K_{2,t} f(x) = \lim_{\varepsilon \to 0+} \int_{y \in O_\varepsilon} \frac{1}{\sqrt{g(\mathcal{F}_t x)}} \frac{\partial e_1}{\partial \nu_{\mathcal{F}_t y}}(\mathcal{F}_t y, \mathcal{F}_t x) f(y) \rho_t(y) \, d\sigma(y).$

Our first claim is that the norm of an operator

(3.8) $\qquad \lim_{\varepsilon \to 0+} \int_{y \in \partial\Omega;\, \varepsilon < r(x,y) < s} \frac{1}{\sqrt{g(\mathcal{F}_t x)}} \frac{\partial e_1}{\partial \nu_{\mathcal{F}_t y}}(\mathcal{F}_t y, \mathcal{F}_t x) f(y) \rho_t(y) \, d\sigma(y)$

could be bounded in $L^p$ or $\hbar^p$ norm by $Cs^\delta$ for some $\delta > 0$ small ($\delta$ and $C$ independent of $t$). Showing this is trivial. It can be established that the kernel of the operator (3.8) is bounded by $Cs^\delta |x-y|^{-(n-1-\alpha+2\delta)}$ by (1.23). However, as it was already shown an operator with such kernel maps $L^p$ to $L^p$ (and $\hbar^p$ to $\hbar^p$) boundedly with a norm depending on $p$ and the constant $Cs^\delta$. The other part of the operator (3.7), i.e., when we subtract (3.8) from (3.7) has much better kernel without singularity. Using the fact that $\nabla_x e_1(x,y)$ is $C^\alpha$ off the diagonal we get

$$\left|\int_{y\in\partial\Omega;r(x,y)\geq s}\frac{1}{\sqrt{g(\mathcal{F}_tx)}}\frac{\partial e_1}{\partial\nu_{\mathcal{F}_ty}}(\mathcal{F}_ty,\mathcal{F}_tx)f(y)\rho_t(y)\,d\sigma(y)-\right.$$
(3.9)
$$\left.-\int_{y\in\partial\Omega;r(x,y)\geq s}\frac{1}{\sqrt{g(x)}}\frac{\partial e_1}{\partial\nu_y}(y,x)f(y)\,d\sigma(y)\right|\leq C_st^\delta\|f\|.$$

Putting these two things together yields

(3.10) $$\|(K_{2,t}-K_2)f\|\leq(Cs^\delta+C_st^\delta)\|f\|.$$

Notice that the second constant $C_s$ depends on chosen $s>0$. Nevertheless, (3.6) follows from (3.10). The proof of the other part of (3.6) for $\mathcal{D}$ is much easier since $x\in\Omega$ means that the integral we have to consider is not singular. We leave it to the reader. □

Once having (3.6) it follows that for $t\in[0,\varepsilon]$, $\varepsilon>0$ small there are constants $C_1,C_2>0$ independent on $t$ such that

(3.11) $$C_1\|f\|_{L^p(\partial\Omega)}\leq\|(\tfrac{1}{2}I+K_t)f\|_{L^p(\partial\Omega)}\leq C_2\|f\|_{L^p(\partial\Omega)}.$$

(This estimate hold also for $\hbar^p$). Take now

(3.12) $$g_n=(\tfrac{1}{2}I+K_{1/n})^{-1}u_{1/n}\big|_{\partial\Omega}.$$

By (3.11) and we can estimate:

(3.13) $$\|g_n\|_{L^p(\partial\Omega)}\leq C\|u_{1/n}\big|_{\partial\Omega}\|_{L^p(\partial\Omega)}\approx\|(u_{1/n})^*\|_{L^p(\partial\Omega)}\leq\|u^*\|_{L^p(\partial\Omega)}.$$

It follows, that the sequence $(g_n)$ is bounded in $L^p$. Since $p>1$, there is a subsequence, that we also denote by $(g_n)$, which is weakly convergent in $L^p(\partial\Omega)$. Let $g$ be the weak limit of $(g_n)$. For $x\in\Omega$ fixed using (3.5) we get

(3.14) $$u(\mathcal{F}_{1/n}x)=\mathcal{D}_{1/n}(g_n)(x).$$

The left hand side of (3.14) converges to $u(x)$ as $n\to\infty$, since $u$ is continuous inside the domain $\Omega$. Meanwhile, due to the weak convergence of $g_n$ and (3.6) we immediately get that the right side of (3.14) converges to $\mathcal{D}g(x)$. Thus,

(3.15) $$u(x)=\mathcal{D}g(x)\qquad\text{for all }x\in\Omega.$$

This means that $u\big|_{\partial\Omega}=(\tfrac{1}{2}I+K)g$. However, we assumed that $u\big|_{\partial\Omega}=0$. This and invertibility of $\tfrac{1}{2}I+K$ gives that $g=0$. By (3.15) then $u=0$ on $\Omega$. This finishes the proof. □

## 3.2. Hardy space boundary data

The existence and uniqueness can be also established for Hardy spaces in a spirit similar to the $L^p$ case.

THEOREM 3.3. *Let $\partial\Omega \in C^{1+\alpha}$ and the metric tensor $g$ on $M$ be of class $C^\alpha$. Given $f \in \hbar^p(\partial\Omega)$, $\max\{\frac{n-1}{n}, \frac{n-1}{\alpha+n-1}\} < p \le 1$ there exists a unique function $u \in C^{1+\alpha}_{\mathrm{loc}}(\Omega)$ satisfying*

(3.16) $$Lu = 0 \text{ in } \Omega, \qquad u^* \in L^p(\partial\Omega), \qquad u\big|_{\partial\Omega} = f \in \hbar^p(\partial\Omega),$$

*the limit on $\partial\Omega$ taken in the nontangential a.e. sense. Moreover $u$ is representable in the form*

(3.17) $$u = \mathcal{D}((\tfrac{1}{2}I + K)^{-1}f) \text{ in } \Omega,$$

*and there is a uniform estimate*

(3.18) $$\|u^*\|_{L^p(\partial\Omega)} \le C_p \|f\|_{\hbar^p(\partial\Omega)}.$$

PROOF. It remains to establish the uniqueness, since the existence of a solution in the form (3.17) satisfying (3.16) follows from results above. Assume therefore again that $u\big|_{\partial\Omega} = 0$. We look at the proof of a uniqueness given above for $L^p$. Instead of (3.12) take now

(3.18) $$f_n = u_{1/n}\big|_{\partial\Omega}.$$

Define a family of maximal operators $u^{*,t}$ as follows.

(3.19) $$u^{*,t}(x) = \sup_{\substack{y \in \gamma(x) \\ y \notin \Omega_t}} |u(y)|, \qquad \text{for } x \in \partial\Omega.$$

Clearly $u^{*,t} \le u^*$ and $u^{*,t} \le u^{*,t'}$ if and only if $0 < t \le t'$. Since $u\big|_{\partial\Omega} = 0$ it follows that $u^{*,t}(x) \to 0$, as $t \to 0+$ for almost every $x \in \partial\Omega$. This and the fact that $u^* \in L^p(\partial\Omega)$ gives us that $\|u^{*,t}\|_{L^p(\partial\Omega)} \to 0$, as $t \to 0+$.

Now, given the definition of $f_n$, we have that $f_n \in \hbar^p(\partial\Omega)$ and

(3.20) $$\|f_n\|_{\hbar^p(\partial\Omega)} \le C \|u^{*,t}\|_{L^p(\partial\Omega)}, \qquad \text{for any } t > 1/n.$$

Hence, as $n \to \infty$ the norm $\|f_n\|_{\hbar^p(\partial\Omega)}$ goes to zero, or equivalently

(3.21) $$f_n \to 0 \text{ in } \hbar^p(\partial\Omega).$$

Now let $g_n = (\tfrac{1}{2}I + K_{1/n})^{-1} f_n$. Using (3.6) for Hardy spaces implies that

(3.22) $$g_n \to g = 0 \text{ in } \hbar^p(\partial\Omega).$$

With this in hand the final point is that

(3.23) $$u(\mathcal{F}_{1/n}x) = \mathcal{D}_{1/n}(g_n)(x) \to \mathcal{D}g(x) = 0, \qquad \text{as } n \to \infty.$$

Really,

(3.24) $$|\mathcal{D}_{1/n}(g_n)(x) - \mathcal{D}g(x)| \le |\mathcal{D}(g_n - g)(x)| + |(\mathcal{D}_{1/n} - \mathcal{D})(g_n)(x)| \le$$
$$\le |\mathcal{D}(g_n - g)(x)| + \|\mathcal{D}(x) - \mathcal{D}_{1/n}(x)\|_{\mathcal{L}(L^p(\partial\Omega), \mathbb{R})} \|g_n\|_{\hbar^p(\partial\Omega)}.$$

Lemma 3.2 guarantees that the second term converges to zero. The first term also goes to zero, since the kernel of $\mathcal{D}$ for fixed $x \in \Omega$ is of the class $C^\alpha$ in the second variable and $g_n - g \to 0$ in $\hbar^p(\partial\Omega)$ by (3.22).

This was the only missing ingredient to establish (3.23). Since the left hand side of (3.23) converges to $u(x)$ we conclude that $u(x) = 0$. □

## 3.3. Hölder space boundary data

Finally, we have similar result in Hölder spaces $C^r$.

THEOREM 3.4. *Let $\partial\Omega \in C^1$ and let the metric tensor on $M$ be of class $C^{1+\alpha}$ for some $\alpha > 0$. Then for any $r \in (0,1)$ and $f \in C^r(\partial\Omega)$, the $L^2$ solution of the Dirichlet problem*

$$(3.25) \qquad Lu = 0 \text{ in } \Omega, \quad u^* \in L^2(\partial\Omega), \quad u\big|_{\partial\Omega} = f,$$

*has the property that*

$$(3.26) \qquad u \in C^r(\overline{\Omega}) \quad \text{and} \quad \|u\|_{C^r(\overline{\Omega})} \leq C\|f\|_{C^r(\partial\Omega)}.$$

*Furthermore, $u = \mathcal{D}h$ in $\Omega$ for some $h \in C^r(\partial\Omega)$ with $\|h\|_{C^r(\partial\Omega)} \approx \|u\|_{C^r(\overline{\Omega})}$.*

PROOF. In the light of Theorems 3.1 and 2.10 we only need to check that

$$(3.27) \qquad \mathcal{D} : C^r(\partial\Omega) \to C^r(\overline{\Omega}) \quad \text{is bounded for any } r \in (0,1).$$

As it is well known, this will follows from the estimate

$$(3.28) \qquad \text{dist}(x, \partial\Omega)^{1-r}|\nabla \mathcal{D} f(x)| \leq C\|f\|_{C^r(\partial\Omega)} \qquad \text{uniformly for } x \in \Omega.$$

This estimate has been proven in [23] for Lipschitz domains, thus our theorem follows. □

REMARK. At this point we would like to make a small remark on a Dirichlet problem

$$(3.29) \qquad Lu = f \text{ on } \Omega, \quad u\big|_{\partial\Omega} = 0,$$

where we assume that $f \in L^p(\Omega)$ for some $p > 1$. Later in Chapter 8, we will need the fact that $u_t = u\big|_{\partial\Omega_t}$ is a uniformly bounded family of functions in $L^p$ and that

$$(3.30) \qquad \|u_t\|_{L^p(\partial\Omega_t)} \leq C\|f\|_{L^p(\Omega)}.$$

Seeing (3.30) is not difficult. Define a new function $F$ on $M$ by extending $f$ onto the whole $M$, i.e.,

$$(3.31) \qquad F(x) = \begin{cases} f(x), & \text{for } x \in \Omega, \\ 0, & \text{otherwise.} \end{cases}$$

Clearly $\|F\|_{L^p(M)} = \|f\|_{L^p(\Omega)}$. Let $U = L^{-1}(F)$, where $L^{-1}$ is defined by (1.9). On $\Omega$ $LU = f$. Also by (1.8) $U \in H^{1,p}(M)$ and therefore $U$ has $L^p$ traces on $\partial\Omega$ and $\partial\Omega_t$ for $t > 0$. Moreover

$$(3.32) \qquad \|U\big|_{\partial\Omega_t}\|_{L^p(\partial\Omega_t)} \leq C\|U\|_{H^{1,p}(M)} \leq C\|f\|_{L^p(\Omega)}.$$

The constant $C$ in (3.32) does not depend on $t > 0$. (3.32) also works for $U\big|_{\partial\Omega}$.

Consider now the following boundary problem

$$(3.33) \qquad Lw = 0 \text{ on } \Omega, \quad w\big|_{\partial\Omega} = -U\big|_{\partial\Omega} \in L^p(\partial\Omega).$$

The boundary problem (3.33) is solvable for all $1 < p < \infty$ by Theorem 3.1. Moreover, this theorem also gives us the following estimate on $w^*$:

$$(3.34) \qquad \|w^*\|_{L^p(\partial\Omega)} \leq C\|U\big|_{\partial\Omega}\|_{L^p(\partial\Omega)} \leq C\|f\|_{L^p(\Omega)}.$$

The last inequality comes from (3.32). It also follows from (3.34) that

(3.35) $$\|w|_{\partial\Omega_t}\|_{L^p(\partial\Omega_t)} \leq C\|w^*\|_{L^p(\partial\Omega)} \leq C\|f\|_{L^p(\Omega)}.$$

Now clearly $u = U + w$ solves (3.29) and the claim (3.30) follows from (3.32) and (3.35). $\square$

CHAPTER 4

# The Neumann Problem

### 4.1. $L^p$ boundary data

In this chapter we look at the Neumann problem. First, we again treat the $L^p$ case.

THEOREM 4.1. *Let $\partial\Omega \in C^1$, $g \in L^p(\partial\Omega)$ and $1 < p < \infty$. If $V = 0$ on $\Omega$ assume also $\int_{\partial\Omega} g\, d\sigma = 0$. Then the Neumann problem*

$$(4.1) \qquad Lu = 0 \text{ in } \Omega, \qquad \partial_\nu u\big|_{\partial\Omega} = g, \qquad (\nabla u)^* \in L^p(\partial\Omega)$$

*has a solution satisfying*

$$(4.2) \qquad \|(\nabla u)^*\|_{L^p(\partial\Omega)} \leq C_p \|g\|_{L^p(\partial\Omega)}.$$

*If $V > 0$ on a set of positive measure in $\Omega$ then the solution $u$ is unique. If $V = 0$ on $\Omega$ then $u$ is unique up to an additive constant. Moreover, $u$ is representable in the form*

$$(4.3) \qquad u = \mathcal{S}((-\tfrac{1}{2}I + K^*)^{-1}g).$$

PROOF. From the results above it follows that $u$ given by (4.3) solves (4.1). The uniqueness for the Neumann problem was established in Proposition 5.5 of [**23**]. □

### 4.2. Hardy space boundary data

Now we turn to the Hardy spaces $\hbar^p(\partial\Omega)$ for $1 - p \geq 0$ small. Recall the setting from the proof of Theorem 3.1, i.e., let $\Theta$ be a smooth vector field on $M$ transversal to $\partial\Omega$ and pointing into $\Omega$. Define $\mathcal{F}_t$, $g_t$, $d\sigma_t$, $\nu_t$ etc. as in that proof. If $u \in C^1_{\text{loc}}(\Omega)$, we say that $\partial_\nu u = g \in \hbar^p(\partial\Omega)$ for some $(n-1)/n < p < 1$, provided

$$(4.4) \qquad \lim_{t \searrow 0} \int_{\partial\Omega} \frac{\partial u_t}{\partial \nu_t} \psi\, d\sigma_t = \int_{\partial\Omega} g\psi\, d\sigma, \qquad \forall \psi \in C^\alpha(\partial\Omega),$$

where $\alpha = (n-1)(p^{-1} - 1) > 0$. Let us also mention that the jump formula (1.15) remains valid since it holds for $p$-atoms.

THEOREM 4.2. *Let $\partial\Omega \in C^1$, $g \in \hbar^p(\partial\Omega)$ and $\max\{\frac{n-1}{n}, \frac{n-1}{\alpha+n-1}\} < p \leq 1$. If $V = 0$ on $\Omega$ assume also $g \in \hbar^p_{\text{at}}(\partial\Omega)$. Then the Neumann problem*

$$(4.5) \qquad Lu = 0 \text{ in } \Omega, \qquad \partial_\nu u\big|_{\partial\Omega} = g, \qquad u \in C^1_{\text{loc}}(\Omega), \qquad (\nabla u)^* \in L^p(\partial\Omega)$$

*has a solution satisfying*

$$(4.6) \qquad \|(\nabla u)^*\|_{L^p(\partial\Omega)} \leq C_p \|g\|_{\hbar^p(\partial\Omega)}.$$

If $V > 0$ on a set of positive measure in $\Omega$, then $u$ is unique. If $V = 0$ on $\Omega$ then $u$ is unique up to an additive constant. Moreover, the solution $u$ is representable in the form

$$u = \mathcal{S}((-\tfrac{1}{2}I + K^*)^{-1}g). \tag{4.7}$$

PROOF. Again by previous results, (4.7) is well-defined, solves (4.5) and satisfies the estimate (4.6). Therefore we can concentrate on the question of uniqueness. Assume therefore that $u$ solves (4.5) with $g = 0$.

For a moment, consider the case when $V = 0$ on $\overline{\Omega}$. Recall that $\rho_t = d\sigma_t/d\sigma$. Using self-explanatory piece of notation emphasizing the dependence on the metric tensor, we claim that

$$\tfrac{\partial u_t}{\partial \nu_t} \in \hbar^p(\partial\Omega, d\sigma_t) \Longrightarrow \tfrac{\partial u_t}{\partial \nu_t}\rho_t \in \hbar^p(\partial\Omega, d\sigma), \quad \forall t > 0, \tag{4.8}$$

and

$$\left\|\tfrac{\partial u_t}{\partial \nu_t}\rho_t\right\|_{\hbar^p(\partial\Omega,d\sigma)} \leq C \left\|\tfrac{\partial u_t}{\partial \nu_t}\right\|_{\hbar^p(\partial\Omega,d\sigma_t)} \leq C\|(\mathrm{grad}_t u_t)^*\|_{L^p(\partial\Omega,\, d\sigma_t)}$$
$$\leq C\|(\nabla u)^*\|_{L^p(\partial\Omega)}, \tag{4.9}$$

uniformly for $t > 0$. The first membership (4.8) together with the second and third estimate in (4.9) follows from [**31**]. The second membership and the first inequality in (4.9) follow from Appendix A of [**23**]. In the case $V > 0$ on a set of positive measure in $\Omega$ all work for $\widetilde{u} = u - \int_\Omega E(.,y)V(y)u(y)\, d\mathrm{Vol}$ and we get similar conclusions.

To finish the proof we need to establish an analogue of Lemma 3.2.

LEMMA 4.3. *For any* $\max\{\tfrac{n-1}{n}, \tfrac{n-1}{\alpha+n-1}\} < p \leq 1$, *the operators* $K_t^*$ *converge to* $K^*$ *and the operators* $\mathcal{S}_t$ *converge to* $\mathcal{S}$ *in the norm as* $t \searrow 0$. *More precisely*

$$\sup_{\|f\|=1} \|K^*f - \rho_t(K_t^* \tfrac{1}{\rho_t} f)\|_{\hbar^p(\partial\Omega)} \to 0 \quad \text{as } t \searrow 0,$$

(4.10)

*and for any* $x \in \Omega$:
$$\sup_{\|f\|=1} |\mathcal{S}f(x) - \mathcal{S}_t(f\tfrac{1}{\rho_t})(x)| \to 0 \quad \text{as } t \searrow 0,$$

$$\sup_{\|f\|=1} |\nabla_x \mathcal{S}f(x) - \nabla_x \mathcal{S}_t(f\tfrac{1}{\rho_t})(x)| \to 0 \quad \text{as } t \searrow 0.$$

PROOF. We recall that (2.52) applies and validates the first line of (4.10). As far as the operators $\mathcal{S}_t$ are concerned, since $\nabla_x E(x,y)$ is of class $C^\alpha$ off the diagonal, and we have

$$(\nabla \mathcal{S})f(x) = \int_M \nabla_x E(x,y)f(y)\, d\sigma(y) \qquad x \in \Omega, \tag{4.11}$$

we will get (4.10) for a fixed $x$, by Hölder continuity of $\nabla_x E(x,y)$. □

Once having Lemma 4.3, let us assume first that $V > 0$ on a set of positive measure in $\Omega$. Take

$$g_n = \frac{\partial u_{1/n}}{\partial \nu_{1/n}}. \tag{4.12}$$

Since $g_n \in C(\overline{\Omega})$, the uniqueness result for (4.1) in $L^2$ gives

(4.13) $$u(\mathcal{F}_{1/n}x) = \mathcal{S}_t((-\tfrac{1}{2}I + K^*_{1/n})^{-1}g_n).$$

Then (4.9) implies that the sequence $(g_n\rho_n)$ is bounded in $h^p(\partial\Omega, d\sigma)$. In fact, if $(\nabla u)^{*,t}$ is defined exactly as in (3.19), we get that $(\nabla u)^{*,t} \to 0$ in $L^p(\partial\Omega)$, as $t \to 0+$. This and (4.9) then gives us that $g_n\rho_n \to 0$ in $\hbar^p(\partial\Omega)$, as $n \to \infty$.

We put $f_n = (-\tfrac{1}{2}I + K^*_{1/n})^{-1}g_n$. By Lemma 4.3 then $f_n\rho_n \to f = 0$ in $\hbar^p(\partial\Omega)$. With this in hand the final point is that

(4.14) $$u(\mathcal{F}_{1/n}x) = \mathcal{S}_{1/n}(f_n)(x) \to \mathcal{S}f(x) = 0, \quad \text{as } n \to \infty.$$

Actually, also

(4.15) $$\nabla u(\mathcal{F}_{1/n}x) = \nabla\mathcal{S}_{1/n}(f_n)(x) \to \nabla\mathcal{S}f(x) = 0, \quad \text{as } n \to \infty.$$

Here (4.14) and (4.15) are established by techniques used in Theorem 3.3 (using Lemma 4.3). We do not repeat these arguments. Hence $u(x) = 0$ for all $x \in \Omega$.

Similarly, if $V = 0$ on $\Omega$ (4.14) does not hold, since (4.13) is no longer true, but (4.15) can still be established. That is, $\nabla u(x) = 0$ at any point $x \in \Omega$. We conclude that $u$ is a constant function. $\square$

## 4.3. Hölder space boundary data

Finally, it is also possible to establish the Neumann problem in Hölder spaces.

THEOREM 4.4. *Assume that the metric tensor on $M$ and the boundary $\partial\Omega$ are of class $C^{1+\alpha}$ for some $\alpha > 0$. Then for any $r \in (0, \alpha)$ and $g \in C^r(\partial\Omega)$, the $L^2$ solution to the Neumann problem*

(4.16) $$Lu = 0 \text{ in } \Omega, \qquad \partial_\nu u\big|_{\partial\Omega} = g$$

*belongs to $C^{1+r}(\overline{\Omega})$. Moreover, the estimate*

(4.17) $$\|u\|_{C^{1+r}(\overline{\Omega})} \leq C\|g\|_{C^r(\partial\Omega)}$$

*holds. If $V > 0$ on a set of positive measure in $\Omega$ then $u$ is unique. If $V = 0$ on $\Omega$ then $u$ is unique up to an additive constant.*

PROOF. As in Theorem 3.4, it remains to show that given the assumptions, the single layer potential $\mathcal{S}$ maps $C^r(\partial\Omega)$ into $C^{1+r}(\overline{\Omega})$ boundedly for any $r \in (0,1)$. This will follow from the estimate

(4.18) $$\operatorname{dist}(x,\partial\Omega)^{1-r}|\nabla^2\mathcal{S}f(x)| \leq C\|f\|_{C^r(\partial\Omega)}, \quad \text{uniformly for } x \in \Omega.$$

To see this we fix a function $f \in C^\alpha$, a point $x \in \overline{\Omega}$ and select $p \in \partial\Omega$ such that $d = \operatorname{dist}(x,\partial\Omega) = \operatorname{dist}(x,d)$. Since, $\nabla(S1) \in L^\infty$, without loss of generality we can assume that $f(p) = 0$. Now, for a large constant $C$ we split the domain of integration into $\{y \in \partial\Omega : \operatorname{dist}(y,p) \leq Cd\}$ and $\{y \in \partial\Omega : \operatorname{dist}(y,p) > Cd\}$. In the first subdomain, the kernel of resulting integral $\nabla^2\mathcal{D}$ (which is essentially $\nabla_x^2 E(x,y)$) can be majorized by $Cd^n$, while in second subdomain we majorize the kernel by $C\operatorname{dist}(y,p)^{-n}$. This works in the present context due to the decomposition (1.21), as well as, the estimates

(4.19) $$(\nabla_z^2 e_0)(z,y) = \mathcal{O}(|z|^{-n}) \quad \text{as } z \to 0 \text{ uniformly in } y,$$

and

(4.20) $$(\nabla_x^2 e_1)(x,y) \leq C_\varepsilon |x-y|^{-(n-1+\varepsilon)},$$

for any $\varepsilon > 0$. (4.20) can be found in [**21**]. Finally, the fact that $|f(y)| \leq C \operatorname{dist}(y,p)^r$ is used. $\square$

CHAPTER 5

# Compactness of Layer Potentials, Part II
# The Dirichlet regularity problem

## 5.1. Preliminaries

In this chapter we return to the topic studied in the first part of Chapter 2 - the compactness of the operator $K$. This time we would like to prove that $K$ is well defined, bounded and compact on $H^{1,p}(\partial\Omega)$, where $H^{1,p}(\partial\Omega)$ is the Sobolev space (Hardy-Sobolev space if $p \leq 1$).

In general, for $(n-1)/n < p \leq 1$ we define the Hardy-Sobolev space $H^{1,p}(\partial\Omega)$ by

$$(5.1) \qquad H^{1,p}(\partial\Omega) = \{f \in \hbar^p(\partial\Omega) : \nabla f \in \hbar^p(\partial\Omega)\}.$$

Here

$$(5.2) \qquad \nabla f = \nabla_T f$$

is the tangential gradient of $f$ (with respect to $\partial\Omega$). By the embedding theorem (which remains to hold even though $p \leq 1$ as can be seen using atoms) in this case

$$(5.3) \qquad f \in H^{1,p}(\partial\Omega) \Longrightarrow f \in L^q(\partial\Omega) \quad \text{for some } q > 1.$$

For $1 < p < \infty$ the Sobolev space $H^{1,p}(\partial\Omega)$ is defined similarly, i.e.,

$$(5.4) \qquad H^{1,p}(\partial\Omega) = \{f \in L^p(\partial\Omega) : \nabla f \in L^p(\partial\Omega)\}.$$

Again (5.4) guarantees that

$$(5.5) \qquad f \in L^q(\partial\Omega), \qquad \text{for } \frac{1}{q} = \frac{1}{p} - \frac{1}{n}.$$

Our approach as before is going to be a direct analysis of the kernel of the operator $K$. From now on, let us assume higher regularity of the metric tensor on $M$; namely we require that $g \in C^{1+\alpha}$ for some $\alpha > 0$. A result from [21] (Propositions 2.5 and 2.8) gives us that in the decomposition (1.21) of the kernel $E(x,y)$ we have for any $\varepsilon > 0$

$$(5.6) \qquad |\nabla_x^j \nabla_y^k e_1(x,y)| \leq C_\varepsilon |x-y|^{-(n-3+j+k+\varepsilon)},$$

for each $j, k \in \{0, 1\}$.

By (5.3), for our range of $p$ Corollary 2.2 applies and gives us that

$$(5.7) \qquad K : H^{1,p}(\partial\Omega) \to L^q(\partial\Omega)$$

is well defined and compact for some $q > p$. Hence, it suffices to deal with the tangential derivatives of $K$; the goal is to show that

$$(5.8) \qquad \nabla K : H^{1,p}(\partial\Omega) \to L^p(\partial\Omega)$$

($L^p$ replaced by $\hbar^p$ when $p < 1$) is well defined and compact. Here $\nabla$ means gradient on $\partial\Omega$.

Fist we look at a very special case, namely we would like to show that

(5.9) $$\nabla K1 \in L^\infty(\partial\Omega).$$

Here 1 represents a constant function equal to 1 on $\partial\Omega$. Recall that from the definition of $K$:

(5.10) $$K1(x) = \text{P.V.} \int_{\partial\Omega} \frac{\partial E}{\partial \nu_y}(x,y)\, d\sigma(y) = \lim_{\varepsilon \to 0} \int_{O_\varepsilon} \frac{\partial E}{\partial \nu_y}(x,y)\, d\sigma(y).$$

Here $O_\varepsilon = \{y \in \partial\Omega; |x - y| > \varepsilon\}$. For a fixed $x \in \partial\Omega$, consider the region $\Omega_\varepsilon = \{y \in \Omega; |x - y| > \varepsilon\}$. Clearly, $\partial\Omega_\varepsilon = O_\varepsilon \cup S_\varepsilon$, where $S_\varepsilon = \{y \in \Omega; |x - y| = \varepsilon\}$. We want to compute

(5.11) $$\lim_{\varepsilon \to 0+} \int_{S_\varepsilon} \frac{\partial E}{\partial \nu_y}(x,y)\, d\sigma(y).$$

It follows from (5.6) that the contribution of the second term (with $e_1$) to (5.11) goes to zero, as $\varepsilon \to 0$.

On the other hand, for $x$ fixed, we can pick coordinates arbitrarily, in particular, we can achieve that $g_{ij}(x) = \delta_{ij}$. Hence near $x$, $g_{ij} \approx \delta_{ij}$ and thus

(5.12) $$\frac{\partial}{\partial \nu_y} \approx \sum_i \frac{y_i - x_i}{|x - y|} \frac{\partial}{\partial y_i}.$$

This makes the first term of (1.35) look like:

(5.13) $$\frac{1}{\sqrt{g(x)}} \frac{\partial}{\partial \nu_y} e_0(x - y, x) = \frac{\partial}{\partial \nu_y} C_n |x - y|^{-(n-2)} \approx K_n |x - y|^{-(n-1)}.$$

Also notice that, for almost every $x \in \partial\Omega$, $S_\varepsilon$ is essentially half of the surface of a sphere centered at $x$ of radius $\varepsilon$. This means that for almost every $x \in \partial\Omega$

(5.14) $$\lim_{\varepsilon \to 0+} \int_{S_\varepsilon} \frac{\partial E}{\partial \nu_y}(x,y)\, d\sigma(y) = \text{const}.$$

A careful computation reveals that this constant is equal to $\frac{1}{2}$. Therefore

(5.15) $$K1(x) = \frac{1}{2} + \lim_{\varepsilon \to 0} \int_{\partial\Omega_\varepsilon} \frac{\partial E}{\partial \nu_y}(x,y)\, d\sigma(y).$$

On each set $\Omega_\varepsilon$ the function $\frac{\partial E}{\partial \nu_y}$ is not singular. Integration by parts and (5.15) yield:

(5.16) $$K1(x) = \frac{1}{2} + \lim_{\varepsilon \to 0} \int_{\Omega_\varepsilon} \Delta_y E(x,y)\, d\text{Vol}(y).$$

Since $\Delta_y E(x,y) - V(y) E(x,y) = L_y E(x,y) = \delta_x(y)$, we get that

(5.17) $$K1(x) = \frac{1}{2} + \int_\Omega V(y) E(x,y)\, d\text{Vol}(y).$$

Thus

(5.18) $$\nabla_x K1(x) = \int_\Omega V(y) \nabla_x E(x,y)\, d\text{Vol}(y).$$

By (1.23), it is clear that (5.18) is bounded. This proves (5.9).

Recall that the operator $K$ could be decomposed into two parts $K_1$ and $K_2$ (see (2.7)). Hence, we can write

$$\nabla K1 = \nabla K_1 1 + \nabla K_2 1. \tag{5.19}$$

The two integral operators on the right hand side must be understood just in a formal sense or as distributions. There is no guarantee that $\nabla K_1 1$ or $\nabla K_2 1$ are in some $L^p(\partial\Omega)$, for $p > 1$. Nevertheless, their sum belongs to $L^\infty(\partial\Omega)$.

## 5.2. Compactness and invertibility of $K$ on Sobolev space $H^{1,p}$

First, concentrate on the remainder $K_2$ where we would like to show the following:

LEMMA 5.1. *Let $\partial\Omega \in C^1$ and $U$ be a small neighborhood of a point $x \in \partial\Omega$. Consider any smooth coordinates on $U$ and decompose the kernel $E(x,y)$ as in (1.35). Let $\operatorname{supp}\psi(x,y) \subset U \times U$. Consider the operator*

$$K_2 f(x) = \lim_{\varepsilon \to 0+} \int_{y \in O_\varepsilon} \psi(x,y) \frac{1}{\sqrt{g(x)}} \frac{\partial e_1}{\partial \nu_y}(y,x) f(y) \, d\sigma(y). \tag{5.20}$$

*We claim that for any $(n-1)/n < p < \infty$ there is $q > p$ and $q > 1$ such that:*
(a) *$K_2$ is compact from $H^{1,p}(\partial\Omega)$ to $L^q(\partial\Omega)$ and*
(b) *the operator $f \mapsto \nabla K_2 f - f \nabla K_2 1$ is compact from $H^{1,p}(\partial\Omega)$ to $L^q(\partial\Omega)$.*

PROOF. Let us remark that the derivative $\nabla$ is taken in the tangential directions to $\partial\Omega$. Part (a) is trivial and follows from Lemma 2.1 and the fact that $H^{1,p}(\partial\Omega) \hookrightarrow L^q(\partial\Omega)$ for some $q > p$, $q > 1$. To simplify our considerations, let us denote by $T_2$ the operator

$$T_2 f = \nabla K_2 f - f \nabla K_2 1. \tag{5.21}$$

Write $K_2$ given by (5.20) in local coordinates on $U$; that is $U \cap \Omega$ can be written as in (2.3). Using the notation $x = (x', x_n)$ there $x' \in \mathbb{R}^{n-1}$ we get:

$$\widetilde{K_2} f(x') = \int_{y' \in \mathbb{R}^{n-1}} \psi((x', \varphi(x')), (y', \varphi(y'))) \frac{1}{\sqrt{g(x', \varphi(x'))}} \times$$
$$\sum_i \left[ \frac{\partial e_1}{\partial y_i}((y', \varphi(y')), (x', \varphi(x'))) \nu^i(y') \right] f(y') \, \widetilde{d\sigma}(y'). \tag{5.22}$$

Here the measure on $\mathbb{R}^{n-1}$ $\widetilde{d\sigma}$ is obtained from measure $d\sigma$ on $\partial\Omega$. Observe that $f \in L^p(\partial\Omega, d\sigma)$ if and only if $f \in L^p(\mathbb{R}^{n-1}, \widetilde{d\sigma})$. Also $\frac{\partial e_1}{\partial y_i}$ means the partial derivative of $e_1(y,x)$ with respect to variable the $y_i$, for $i = 1, 2, \ldots, n$, and $\nu^i$ is the $i$-th component of the outer normal to $\partial\Omega$.

As we mentioned above, we know that if $f \in H^{1,p}$, then by a Sobolev embedding theorem $f \in L^q$ with $q > p$ and $q > 1$. Pick $q'$ such that

$$q > q' > p \quad \text{and} \quad q' > 1. \tag{5.23}$$

Then there is $\varepsilon > 0$ such that $f \in H^{\varepsilon,q'}$, again by the embedding theorem. Pick any unit vector $e \in \mathbb{R}^{n-1}$. We would like to show that the operator

(5.24)
$$\frac{\partial}{\partial e}K_2 f(x') - f(x')\frac{\partial}{\partial e}K_2 1(x')$$
$$= \lim_{h \to 0} \frac{\overline{K_2}f(x'+he) - \overline{K_2}f(x') - f(x')\overline{K_2}1(x'+he) + f(x')\overline{K_2}1(x')}{h}$$

maps $H^{\varepsilon,q'}$ onto $L^{q'}$ boundedly. This would guarantee that the operator $T_2$ mapping $H^{1,p}(\partial\Omega)$ into $L^{q'}(\partial\Omega)$ is bounded and compact since the embedding of $H^{1,p}(\partial\Omega)$ into $H^{\varepsilon,q'}(\partial\Omega)$ is compact.

Evaluating (5.24) using (5.22) we basically get that (5.24) can be written as sum of two integrals. The first one has domain of integration $|x'-y'| < \eta$ and the other one $|x'-y'| \geq \eta$. The first integral goes to zero as $\eta \to 0$ and the second integral as $\eta \to 0$ converges to

(5.25)
$$\int_{y' \in \mathbb{R}^{n-1}} \frac{\partial}{\partial e}\left[\psi((x',\varphi(x')),(y',\varphi(y')))\frac{1}{\sqrt{g(x',\varphi(x'))}}\right] \times$$
$$\sum_i \left[\frac{\partial e_1}{\partial y_i}((y',\varphi(y')),(x',\varphi(x')))\nu^i(y')\right] f(y') \widetilde{d\sigma}(y') -$$
$$- \int_{y' \in \mathbb{R}^{n-1}} \frac{\partial}{\partial e}\left[\psi((x',\varphi(x')),(y',\varphi(y')))\frac{1}{\sqrt{g(x',\varphi(x'))}}\right] \times$$
$$\sum_i \left[\frac{\partial e_1}{\partial y_i}((y',\varphi(y')),(x',\varphi(x')))\nu^i(y')\right] f(x') \widetilde{d\sigma}(y') +$$
$$+ \int_{y' \in \mathbb{R}^{n-1}} \psi((x',\varphi(x')),(y',\varphi(y')))\frac{1}{\sqrt{g(x',\varphi(x'))}} \times$$
$$\sum_i \left[\frac{\partial^2 e_1}{\partial e \partial y_i}((y',\varphi(y')),(x',\varphi(x')))\nu^i(y')\right] [f(y')-f(x')] \widetilde{d\sigma}(y').$$

The first two integrals apparently maps $L^{q'}(\partial\Omega)$ onto $L^{q'}(\partial\Omega)$ boundedly. The arguments is essentially same as in Corollary 2.2. On the other hand the kernel of the third integral $K(x,y)$ can be estimated using (5.6) which leads to

(5.26)
$$|K(x',y')| \leq C|x'-y'|^{-(n-1+\delta)},$$

for any $\delta > 0$. Also $K$ is continuous off the diagonal $\{x'=y'\}$. Now Proposition A.17 gives that any operator with such kernel maps $H^{\varepsilon,q'}(\partial\Omega)$ to $L^{q'}(\partial\Omega)$. $\square$

Consider now the same sequence of domains $\Omega_1 \subset \Omega_2 \subset \dots$ approximating $\Omega = \Omega_0$ as in Chapter 2, i.e., we have (2.3) together with (2.4). By $K_k = K_{k,1} + K_{k,2}$, for $k = 0, 1, 2, \dots$, we again denote the corresponding boundary operator to $\partial\Omega_k$. Consider also the operators

(5.27)
$$T_k f = \nabla K_k f - f \nabla K_k 1 = T_{k,1} f + T_{k,2} f,$$

where $T_{k,1}$, $T_{k,2}$ are pieces of the operator $T_k$ that corresponds to kernel decomposition (1.35).

We can see that $T_{k,1}f$ are well defined and map $H^{1,p}(\partial\Omega)$ into $L^p(\partial\Omega)$ (c.f. Proposition A.10). Indeed, if we look for example at the operator $K_1 = K_{0,1}$ we get

$$K_1 f(x) = \lim_{\varepsilon \to 0+}$$
(5.28)
$$\int_{y \in O_\varepsilon} \frac{K_n \psi(x,y)}{\sqrt{g(x)}} \left( \frac{\sum_i \nu^i(y) \sum_j g_{ij}(x)(x_j - y_j)}{\left(\sum_{jk} g_{jk}(x)(x_j - y_j)(x_k - y_k)\right)^{n/2}} \right) f(y) \, d\sigma(y).$$

Now writing $x$ as $(x', \varphi(x'))$, where $x' \in \mathbb{R}^{n-1}$ we can formally compute $\frac{\partial}{\partial_i} K_1 f - f \frac{\partial}{\partial_i} K_1 1$ for $i = 1, 2, \ldots, n-1$. Looking at (5.28) we conclude that we get integrals with kernels of three different types. The mentioned integrals are

$$\int_\Gamma b(x, x-y) g(y) f(y) \, d\sigma(y), \quad b(x,z) \text{ odd and homog. of degree } -(n-1) \text{ in } z$$

$$\int_\Gamma b(x, x-y) g(y) f(x) \, d\sigma(y), \quad b(x,z) \text{ odd and homog. of degree } -(n-1) \text{ in } z$$

(5.29)
$$\int_\Gamma b(x, x-y) g(y) [f(y) - f(x)] \, d\sigma(y), \quad b(x,z) \text{ even and homog. of deg. } -n \text{ in } z.$$

Here, since $\psi(x,y)$ can be decomposed as $\phi^i(x)\phi^j(y)$, $g(y) = \psi^j(y)\nu^k(y)$ is a continuous function. Because $L^p$, $p > 1$ is a module over continuous functions we can include $g$ into $f$ and Proposition A.4. takes care of the first two types of integrals. For the third type we apply Proposition A.10. These two Propositions also gives us that any slight $C^1$ alteration of the boundary $\Gamma$ does not change the resulting integral much (in the norm). In particular our sequence of operators $T_{k,1}$ converges to $T_{0,1} = T_1$ as $k \to \infty$.

If we put together this and Lemma 5.1, we can conclude that for any $1 < p < \infty$ the operators $T_k$, $k = 0, 1, 2, \ldots$ map $H^{1,p}(\partial\Omega)$ to $L^p(\partial\Omega)$.

Finally, we briefly discuss the compactness of the operator $T_k$, $k \geq 1$. As we saw above, for each $k = 1, 2, \ldots$ we can choose local coordinates such that the first part $T_{k,1}$ of the operator $T_k$ is zero, provided the considered metric tensor on $M$ is *smooth*. Thus by Lemma 5.1, $T_k$ is compact in this case.

If $g$ the metric tensor on $M$ is not *smooth* we can still do what we did in Chapter 2, i.e., approximate $g$ by a sequence $g^\mu$ of *smooth* metric tensors such that

(5.30) $\quad g^\mu \to g \quad$ uniformly in $C^{1+\gamma}$ for any $\gamma < \alpha$ on $M$ as $\mu \to \infty$.

Then arguing exactly as in Chapter 2 and using Proposition A.10 we get that we can make the operator $T_{k,1}$ arbitrary small in its norm. This and the compactness of $T_{k,2}$ from Lemma 5.1 yield that for each $k = 1, 2, \ldots$ the operator $T_k$ is compact from $H^{1,p}(\partial\Omega)$ to $L^p(\partial\Omega)$.

We summarize this result in the following Proposition.

PROPOSITION 5.2. *Let $\partial\Omega \in C^1$ and $1 < p < \infty$. Then the operator*

(5.31) $$K : H^{1,p}(\partial\Omega) \to H^{1,p}(\partial\Omega)$$

*is well defined, bounded and compact.*

PROOF. We know that the operators $T_{k,1}$ are compact for $k = 1, 2, \ldots$. Since

(5.32) $$\|T_{k,1} - T_1\|_{\mathcal{L}(H^{1,p}(\partial\Omega), L^p(\partial\Omega))} \to 0, \quad \text{as } k \to \infty,$$

we have that $T_1$ is compact. Then, by Lemma 5.1 $T = T_1 + T_2$ must be compact. Now, the operator $f \mapsto f\nabla K1$ from $H^{1,p}(\partial\Omega_k)$ to $L^p(\partial\Omega_k)$ is compact, since $f \in H^{1,p}(\partial\Omega)$ and $\nabla K1 \in L^\infty(\partial\Omega)$. Finally, $\nabla Kf = Tf + f\nabla K1$ and hence $\nabla K$ is compact from $H^{1,p}(\partial\Omega)$ to $L^p(\partial\Omega)$. □

Now exactly as in Chapter 2 we can also prove the following.

PROPOSITION 5.3. *Let $1 < p < \infty$. Then the operator*

(5.33) $$\tfrac{1}{2}I + K : H^{1,p}(\partial\Omega) \to H^{1,p}(\partial\Omega)$$

*is well defined, bounded and invertible.*

PROOF. Clearly (5.33) is Fredholm. It also has index zero by the same argument as in Chapter 2. □

Now we would like to establish following result which will be needed in next chapter.

LEMMA 5.4. *Let $1 < p < \infty$. Consider the double layer potential $\mathcal{D}$ on $H^{1,p}(\partial\Omega)$. There exists a constant $C = C(\partial\Omega, p)$ such that*

(5.34) $$\|(\nabla \mathcal{D}f)^*\|_{L^p(\partial\Omega)} \leq C\|f\|_{H^{1,p}(\partial\Omega)}.$$

PROOF. Once again we use (1.35). It follows, that we can write operator $\mathcal{D}$ as $\mathcal{D} = \mathcal{D}_1 + \mathcal{D}_2$. For $x \in \Omega$ and $x_0 \in \partial\Omega$ we put:

(5.35)
$$\begin{aligned}\mathcal{R}f(x,x_0) &= \mathcal{R}_1 f(x,x_0) + \mathcal{R}_2 f(x,x_0) = \\ &= (\nabla \mathcal{D}_1 f(x) - f(x_0)\nabla \mathcal{D}_1 1(x)) + (\nabla \mathcal{D}_2 f(x) - f(x_0)\nabla \mathcal{D}_2 1(x)).\end{aligned}$$

The dealing with $\mathcal{R}_2$, with kernel $\frac{1}{\sqrt{g(x)}}e_1(y,x)$, is very similar to analysis in the proof of Lemma 5.1 for the part $T_2$ of the operator $T$. We therefore skip this step and just state the result:

(5.36) $$\|\sup_{x \in \gamma(x_0)} |\mathcal{R}_2 f(x,x_0)|\|_{L^p(\partial\Omega)} \leq C\|f\|_{H^{1,p}(\partial\Omega)}.$$

## 5.2. COMPACTNESS AND INVERTIBILITY OF $K$ ON SOBOLEV SPACE $H^{1,p}$

Now, we consider the maximal operator $\mathcal{R}_1$. Written in local coordinates the kernel of $\frac{\partial}{\partial x_l}\mathcal{D}_1$, $l = 1, 2, \ldots, n$ looks like

$$\frac{\partial}{\partial x_l}\left(\frac{K_n\psi(x,y)}{\sqrt{g(x)}}\right)\left(\sum_i \frac{\nu^i(y)\sum_j g_{ij}(x)(x_j - y_j)}{\left(\sum_{jk} g_{jk}(x)(x_j - y_j)(x_k - y_k)\right)^{n/2}}\right)+$$

$$+\frac{K_n\psi(x,y)}{\sqrt{g(x)}}\left(\frac{\nu^l(y)\sum_j g_{lj}(x)}{\left(\sum_{jk} g_{jk}(x)(x_j - y_j)(x_k - y_k)\right)^{n/2}}\right)+$$

(5.37)

$$+\frac{K_n\psi(x,y)}{\sqrt{g(x)}}\left(\sum_i \frac{\nu^i(y)\sum_j \frac{\partial}{\partial x_l}(g_{ij}(x))(x_j - y_j)}{\left(\sum_{jk} g_{jk}(x)(x_j - y_j)(x_k - y_k)\right)^{n/2}}\right)+$$

$$+\frac{\widetilde{K_n}\psi(x,y)}{\sqrt{g(x)}}\left(\sum_i \frac{\nu^i(y)\sum_{j,k} g_{ij}(x)g_{lk}(x)(x_j - y_j)(x_k - y_k)}{\left(\sum_{jk} g_{jk}(x)(x_j - y_j)(x_k - y_k)\right)^{(n+2)/2}}\right)+$$

$$+\frac{\widetilde{K_n}\psi(x,y)}{\sqrt{g(x)}}\left(\sum_i \frac{\nu^i(y)\sum_{j,k,t} g_{ij}(x)\frac{\partial}{\partial x_l}(g_{kt}(x))(x_j - y_j)(x_k - y_k)(x_t - y_t)}{\left(\sum_{jk} g_{jk}(x)(x_j - y_j)(x_k - y_k)\right)^{(n+2)/2}}\right).$$

Hence, to estimate $\mathcal{R}_1(x, x_0)$ we apply Proposition 1.5 of [**22**] to the first, third and fifth terms in (5.37). We conclude that a maximal operator with these kernels is bounded by $C\|f\|_{L^p(\partial\Omega)}$. We deal with the other two terms by invoking Proposition A.12 of the appendix. Both are of the form $b(x - y, x)g(y)$, where the function $b$ is even and homogeneous of order $-n$ in $x - y$, smooth enough in this variable and the function $g$ is continuous. Thus according to Proposition A.12, for an operator $\mathcal{B}$

(5.38) $$\mathcal{B}f(x, x_0) = \int_\Gamma b(x, x - y)g(y)\left[f(y) - f(x_0)\right]\, d\sigma(y),$$

where $x \notin \Gamma$, $x_0 \in \Gamma$ we have following estimate on nontangential maximal function:

(5.39) $$\|\sup_{x \in \gamma(x_0)} |\mathcal{B}f(x, x_0)|\|_{L^p(\partial\Omega)} \leq C\|g\|_{L^\infty}\|f\|_{H^{1,p}(\partial\Omega)},$$

where $\gamma(x_0)$ in nontangential approach region to the point $x_0$. If we put (5.36) and (5.39) together, we get:

(5.40) $$\|\sup_{x \in \gamma(x_0)} |\mathcal{R}f(x, x_0)|\|_{L^p(\partial\Omega)} \leq C\|f\|_{H^{1,p}(\partial\Omega)}.$$

Recall the relation between $\mathcal{D}$ and $\mathcal{R}$. Let

(5.41) $$\mathcal{T}f(x, x_0) = \nabla\mathcal{D}f(x) - \mathcal{R}(x, x_0) = f(x_0)\nabla\mathcal{D}1(x).$$

The claim is that we have an estimate of the form (5.40) for $\mathcal{T}$. If this claim is true, then for $\nabla \mathcal{D}$ we get:

$$\| \sup_{x \in \gamma(x_0)} |\nabla \mathcal{D}f(x)| \|_{L^p(\partial\Omega)} \leq \tag{5.42}$$
$$\leq C \| \sup_{x \in \gamma(x_0)} |\mathcal{R}f(x,x_0)| \|_{L^p(\partial\Omega)} + C \| \sup_{x \in \gamma(x_0)} |\mathcal{T}f(x,x_0)| \|_{L^p(\partial\Omega)} \leq C \|f\|_{H^{1,p}(\partial\Omega)},$$

so (5.34) is indeed true. Hence, we only need to show that

$$(\nabla \mathcal{D}1)^* \in L^\infty(\partial\Omega). \tag{5.43}$$

However, this is easy. For $x$ fixed, we consider the regions $\Omega_\varepsilon = \Omega \setminus B(x,\varepsilon)$, where $B(x,\varepsilon)$ is a ball of radius $\varepsilon$ centered at $x$ in some smooth coordinate system near $x$. Clearly,

$$\mathcal{D}1(x) = \int_{\partial\Omega} \frac{\partial E}{\partial \nu_y}(x,y) \, d\sigma(y), \tag{5.44}$$

and therefore integrating by parts we get:

$$\mathcal{D}1(x) = \int_{\Omega_\varepsilon} \Delta_y E(x,y) \, d\text{Vol}(y) + \int_{\partial B(x,\varepsilon)} \frac{\partial E}{\partial \nu_y}(x,y) \, d\sigma(y). \tag{5.45}$$

If we limit $\varepsilon \to 0+$, we get that the second (boundary) integral on the right hand side of (5.45) converges to some constant (actually 1), independent of $x$. The argument is same as in the proof of (5.14). On the other hand, $\Delta_y E(x,y) = L_y E(x,y) + V(y)E(x,y)$ and $L_y E(x,y) = \delta_x(y)$. Hence:

$$\mathcal{D}1(x) = 1 + \int_\Omega V(y)E(x,y) \, d\text{Vol}(y). \tag{5.46}$$

From this $\nabla \mathcal{D}1 \in L^\infty(\Omega)$, so (5.43) holds. $\square$

## 5.3. Compactness and invertibility of $K$ on Hardy-Sobolev space $H^{1,p}$

Now we look at a result similar to Lemma 5.4 for Hardy spaces.

LEMMA 5.5. *Let $\partial\Omega \in C^1$ and $(n-1)/n < p \leq 1$. Consider the double layer potential $\mathcal{D}$ on $H^{1,p}(\partial\Omega)$. There exists a constant $C = C(\partial\Omega, p)$ such that*

$$\|(\nabla \mathcal{D}f)^*\|_{L^p(\partial\Omega)} \leq C\|f\|_{H^{1,p}(\partial\Omega)}. \tag{5.47}$$

PROOF. This proof is very similar to the previous one, so we will be brief. Again we can write the operator $\mathcal{D}$ as $\mathcal{D}_1 + \mathcal{D}_2$ and also define the operator $\mathcal{R} = \mathcal{R}_1 + \mathcal{R}_2$. Estimates for part $\mathcal{R}_2$ are again easy, using that $f \in H^{1,p} \hookrightarrow L^q$ for some $q > 1$. The kernel of $\mathcal{D}_1$ could again be written as (5.37) with first, third and fifth integrals easily estimable using Proposition 1.5 of [22]. The remaining terms have again the form $b(x-y,x)g(y)$, where the function $b$ is even and homogeneous of order $-n$ in $x-y$, smooth enough in this variable and the function $g$ is continuous. Thus according to Proposition A.16 for any operator $\mathcal{B}$

$$\mathcal{B}f(x,x_0) = \int_\Gamma b(x, x-y)g(y)\left[f(y) - f(x_0)\right] d\sigma(y), \tag{5.48}$$

## 5.3. COMPACTNESS AND INVERTIBILITY OF $K$ ON HARDY-SOBOLEV SPACE $H^{1,p}$

where $x \notin \Gamma$, $x_0 \in \Gamma$, we have the following estimate on the nontangential maximal function:

(5.49) $$\|\sup_{x \in \gamma(x_0)} |\mathcal{B}f(x,x_0)|\|_{L^p(\partial\Omega)} \leq C\|g\|_{L^\infty}\|f\|_{H^{1,p}(\partial\Omega)},$$

where $\gamma(x_0)$ is a nontangential approach region to the point $x_0$. Now, (5.49) easily yields (5.47) exactly as in Lemma 5.4. □

PROPOSITION 5.6. *Let $\partial\Omega \in C^1$ and $(n-1)/n < p \leq 1$. Then the operator*

(5.50) $$K : H^{1,p}(\partial\Omega) \to H^{1,p}(\partial\Omega)$$

*is well defined, bounded and compact.*

PROOF. We first show that (5.50) is well defined. To see this pick $f \in H^{1,p}(\partial\Omega)$ and consider $u = \mathcal{D}f$. Assume first that $V = 0$ on $\Omega$, i.e., function $u$ is harmonic. Lemma 5.5 yields that

(5.51) $$\|(\nabla u)^*\|_{L^p(\partial\Omega)} \leq C\|f\|_{H^{1,p}(\partial\Omega)}.$$

As follows from [**31**], (5.51) gives $\partial_\nu u \in \hbar^p(\partial\Omega)$ and $\nabla_T u \in \hbar^p(\partial\Omega)$, $\nabla_T$ meaning gradient on $\partial\Omega$. However, $u|_{\partial\Omega} = (\frac{1}{2}I + K)f$ which gives
(5.52)
$$\|\nabla_T((\tfrac{1}{2}I + K)f)\|_{\hbar^p(\partial\Omega)} = \|\nabla_T u\|_{\hbar^p(\partial\Omega)} \leq C\|(\nabla u)^*\|_{L^p(\partial\Omega)} \leq C\|f\|_{H^{1,p}(\partial\Omega)}.$$

From (5.52) it follows that $\nabla_T K$ maps $H^{1,p}(\partial\Omega)$ into $\hbar^p(\partial\Omega)$, hence (5.50) is well defined and bounded.

We turn to compactness now. Take again a sequence of smooth domains $\Omega_1 \subset \Omega_2 \subset \ldots$ approximating $\Omega$ for which in local coordinates we have (2.3) and (2.4). Also as before, let $K_k$ be the operator (5.50) corresponding to the domain $\Omega_k$.

We already know that each $K_k$ is compact for $k = 1, 2, \ldots$, provided the considered metric tensor $g$ on $M$ is *smooth*. If this is not the case, there is a sequence of *smooth* tensors $g^\mu$ already considered earlier in this chapter that approximates $g$ well. An important point here is that although the Hardy space $\hbar^p(\partial\Omega_k)$ in general depends on the metric on $\partial\Omega_k$, all metrics considered here (i.e. $g^\mu$ and $g$) generate the same Hardy space.

Estimating the difference $v_k^\mu = u - u^\mu = (\mathcal{D}_k - \mathcal{D}_k^\mu)f$ (by Proposition A.16), we get that $(\nabla v_k^\mu)^* \to 0$ in $L^p(\partial\Omega_k)$ as $\mu \to \infty$. From that it follows (as in (5.52)) that

$$\|\nabla_T((K_k - K_k^\mu)f)\|_{\hbar^p(\partial\Omega_k)} \leq \|\nabla_T((\tfrac{1}{2}I + K_k - \tfrac{1}{2}I - K_k^\mu)f)\|_{\hbar^p(\partial\Omega_k)} \leq$$
(5.53)
$$\leq C\|(\nabla v_k^\mu)^*\|_{L^p(\partial\Omega_k)} \to 0.$$

So the compactness of $K_k$ follows, since $K_k^\mu \to K_k$ in $\mathcal{L}(H^{1,p})$ norm as $\mu \to \infty$.

Consider again $\Theta$ a smooth vector field on $M$ transversal to $\partial\Omega$ which allows us via its flow to identify $\partial\Omega_k$ with $\partial\Omega$. That is, we can think about $f \in H^{1,p}(\partial\Omega_k)$ also as a function from $f \in H^{1,p}(\partial\Omega_k, d\sigma_k)$, where $d\sigma_k$ is the pull-back of the surface measure from $\partial\Omega_k$ onto $\partial\Omega$. In the case of $C^1$ domains, the measures $d\sigma$ and $d\sigma_k$ are mutually absolutely continuous since the Radon-Nikodym derivative $\rho_k = \frac{d\sigma_k}{d\sigma}$ is a continuous function and $\rho_k \to 1$ uniformly in $C(\partial\Omega)$, as $k \to \infty$. Also the norms of $f$ on $H^{1,p}(\partial\Omega_k, d\sigma_k)$ are comparable with the norm $H^{1,p}(\partial\Omega, d\sigma)$.

From now on, the proof somehow resembles the proof of Proposition 2.7. Take again for simplicity just the case $V = 0$ on $\Omega$. Pick $f \in H^{1,p}(\partial\Omega)$ of norm one and let $u = \mathcal{D}f$. Then $u$ is harmonic and as in (5.51) we have

$$(5.54) \qquad \|(\nabla u)^*\|_{L^p(\partial\Omega)} \leq C\|f\|_{H^{1,p}(\partial\Omega)},$$

and also for $u_k = \mathcal{D}_k f$ ($\mathcal{D}_k$ is the appropriate operator on $\Omega_k$)

$$(5.55) \qquad \|(\nabla u_k)^*\|_{L^p(\partial\Omega_k)} \leq C\|f\|_{H^{1,p}(\partial\Omega_k)},$$

with $C$ independent of $k$.

By Proposition A.16 (and the partition of unity), the contribution of the main piece $\nabla D_1$ can be estimated quite easily. The remainder $\nabla D_2$ can be estimated using (5.6). This yields

$$(5.56) \qquad \|(\nabla(\mathcal{D}_k f - \mathcal{D}f))^*\|_{L^p(\partial\Omega)} \leq \varepsilon_k, \qquad \text{as } k \to \infty.$$

Here $\varepsilon_k \searrow 0$, as $k \to \infty$ is a sequence independent of $f$.

In particular, if we take the tangential derivative $\nabla_T = \frac{\partial}{\partial T}$ we get

$$(5.57) \qquad \|\nabla_T(\mathcal{D}_k f - \mathcal{D}f))^*\|_{L^p(\partial\Omega)} \leq \varepsilon_k, \qquad \text{as } k \to \infty.$$

The meaning of (5.57) should be clarified. With a point $x_0 \in \partial\Omega$ fixed, $\nabla_T(x_0)$ is well defined at $x_0$. We extend it by parallel transport to a neighborhood $U$ of $x_0$. This means

$$(5.58) \qquad \nabla_T(\mathcal{D}_k f - \mathcal{D}f)$$

is now well defined on $\Omega \cap U$ and thus we can evaluate the maximal operator in (5.56). Again, as it is explained in Appendix A, for $x_0 \in \partial\Omega$ there is exactly one point $x_0' \in \Omega_k$ for which

$$(5.59) \qquad x_0' = \mathcal{F}_t x_0 \qquad \text{for some } t > 0,$$

where $\mathcal{F}_t$ is the flow generated by $\Theta$. Thus (5.56) is understood in a sense that we pull back the whole domain of the operator $\mathcal{D}_k(f)$ (via the flow $\mathcal{F}_t$), such that points $x_0$ and $x_0'$ coincide and *then* compute the difference (5.58) and from it the maximal operator.

If $\nabla_{T_k}$ means the tangent derivative with respect to $\partial\Omega_k$, the assumption about the domains $\Omega_k$ gives us that we can write

$$(5.60) \qquad \nabla_{T_k} = \mathcal{A}_k \nabla_T + \mathcal{B}_k \partial \nu_k,$$

where $\partial \nu_k$ means the normal derivative with respect to $\partial\Omega_k$, $\mathcal{A}_k$ is a real and $\mathcal{B}_k$ a vector valued function. Moreover, $\mathcal{A}_k \to 1$ and $\mathcal{B}_k \to 0$ in the $L^\infty$ norm, as $k \to \infty$.

Combining (2.50) with (5.58) and the fact that from (5.57) we get $L^p$ uniform bound on $(\nabla \mathcal{D}_k(\mathcal{B}_k f))^*$ we conclude exactly as in Proposition 2.7:

$$(5.61) \qquad \|(\nabla_{T_k}(\mathcal{D}_k f) - \nabla_T(\mathcal{D}f))^*\|_{L^p(\partial\Omega)} \leq \varepsilon_k', \qquad \text{as } k \to \infty,$$

for a sequence $(\varepsilon_k')_{n \in \mathbb{N}}$ converging to zero.

Hence due to result of Wilson [**31**] as before it follows that

$$(5.62) \qquad \|\tfrac{1}{\rho_k}\nabla_{T_k}(\mathcal{D}_k f) - \nabla_T(\mathcal{D}f)\|_{h^p(\partial\Omega)} \leq \varepsilon_k'', \qquad \text{as } k \to \infty.$$

Finally, this gives

$$\|\tfrac{1}{\rho_k}\nabla_{T_k}K_kf - \nabla_T Kf\|_{h^p(\partial\Omega)} \approx \|\tfrac{1}{\rho_k}\nabla_{T_k}(\tfrac{1}{2}I+K_k)f - \nabla_T(\tfrac{1}{2}I+K)f\|_{h^p(\partial\Omega)} =$$
(5.63)
$$= \|\tfrac{1}{\rho_k}\nabla_{T_k}(\mathcal{D}_kf) - \nabla_T(\mathcal{D}f)\|_{h^p(\partial\Omega)} \leq \varepsilon_k'' \to 0.$$

In the first line of (5.63) we used a simple observation

(5.64) $$\|\tfrac{1}{\rho_k}\nabla_{T_k}f - \nabla_T f\|_{h^p(\partial\Omega)} \to 0.$$

(5.63) is the desired result. Now the compactness of $\nabla_{T_k}K_k$ follows from compactness of $\nabla_T K$. □

As a corollary we get:

PROPOSITION 5.7. *Let $(n-1)/n < p \leq 1$. Then the operator*

(5.65) $$\tfrac{1}{2}I + K : H^{1,p}(\partial\Omega) \to H^{1,p}(\partial\Omega)$$

*is well defined, bounded and invertible.*

## 5.4 Dirichlet regularity problem, Sobolev $H^{1,p}$ ($1 < p < \infty$) data

In this section we establish the existence of a solution to the Dirichlet problem with boundary data in $H^{1,p}$. The spirit of our argument is very similar to Chapters 3 and 4.

THEOREM 5.8. *Assume $\partial\Omega \in C^1$. Let the metric tensor on $M$ be of class $C^{1+\alpha}$. Given $f \in H^{1,p}(\partial\Omega)$, $1 < p < \infty$ there exists a unique function $u \in C^{2+\alpha}_{\text{loc}}(\Omega)$ satisfying*

(5.66) $$Lu = 0 \ in \ \Omega, \qquad (\nabla u)^* \in L^p(\partial\Omega), \qquad u\big|_{\partial\Omega} = f,$$

*the limit on $\partial\Omega$ taken in the nontangential a.e. sense. Moreover $u$ is representable in the form*

(5.67) $$u = \mathcal{D}((\tfrac{1}{2}I+K)^{-1}f) \ in \ \Omega,$$

*and there is a uniform estimate*

(5.68) $$\|(\nabla u)^*\|_{L^p(\partial\Omega)} \leq C_p\|f\|_{H^{1,p}(\partial\Omega)}.$$

REMARK. Using different methods it has been shown in [**24**] that in the range $1-\varepsilon < p < 2+\varepsilon$ for some $\varepsilon = \varepsilon(\partial\Omega) > 0$ the proposition above can be established even for the metric tensor of class $C^\alpha$. Although we believe that same should be true on $C^1$ domains for all $1 < p < \infty$, major technical difficulties prevented us from establishing this result. In particular, the analysis presented in the previous chapter becomes much more complicated. Let us also remark that it follows from (5.68) that $u \in H^{1,q}(\Omega)$, where $q = pn/(n-1)$.

PROOF. Proposition 5.3 and Lemma 5.4 give us that $u$ of the form (5.67) solves (5.66) and satisfies (5.68). Uniqueness follows from uniqueness for Dirichlet problem in Theorem 3.1. □

COROLLARY 5.9. *The operator*

(5.69) $$S : L^p(\partial\Omega) \to H^{1,p}(\partial\Omega)$$

*is invertible for $1 < p < \infty$. In particular the solution to (5.66) can be also written as a single layer potential*

(5.70) $$u = \mathcal{S}(S^{-1}f).$$

PROOF. Pick any $f \in H^{1,p}$ and assume that $u$ solves (5.66). In particular, we have $\partial_\nu u \in L^p$. Using results from Chapter 4 for the Neumann problem, there exists $g \in L^p(\partial\Omega)$ such that

(5.71) $$u = \mathcal{S}g.$$

Our claim is that $f = Sg$. Seeing this is actually not difficult, since $f = u|_{\partial\Omega} = \mathcal{S}_+ g = Sg$. It follows, that the range of the map (5.69) is the whole space $H^{1,p}$. Now we argue that the kernel of $S$ contains only zero. Assume that $Sg = 0$. Then $u = \mathcal{S}g$ solves (5.66) for $f = Sg = 0$. Uniqueness in Theorem 5.8 guarantees that such $u$ is identically zero. From that we have

(5.72) $$0 = \partial_\nu u = \left(\frac{\partial \mathcal{S}g}{\partial \nu}\right)_+ = (-\tfrac{1}{2}I + K^*)g.$$

Clearly, if $V > 0$ on set of positive measure this immediately gives $g = 0$ since the operator $-\tfrac{1}{2}I + K^*$ is invertible. If $V = 0$ on $\Omega$ this implies $g = const$. If the constant is not zero, it would follow that $f = Sg = const \neq 0$ which is contradiction. Therefore $g = 0$.

This proves that operator (5.69) is invertible, since it is injective and its range is the whole target space. (5.70) also follows. $\square$

## 5.5. Dirichlet regularity problem, $H^{1,p}$ $((n-1)/n < p \leq 1)$ data

In a spirit similar to the result above we also have:

THEOREM 5.10. *Assume $\partial\Omega \in C^1$ and $(n-1)/n < p \leq 1$. Let the metric tensor on $M$ be of class $C^{1+\alpha}$. Given $f \in H^{1,p}(\partial\Omega)$, there exists a unique function $u \in C^{2+\alpha}_{loc}(\Omega)$ satisfying*

(5.73) $$Lu = 0 \text{ in } \Omega, \quad (\nabla u)^* \in L^p(\partial\Omega), \quad u|_{\partial\Omega} = f,$$

*the limit on $\partial\Omega$ taken in the nontangential a.e. sense. Moreover, $u$ is representable in the form*

(5.74) $$u = \mathcal{D}((\tfrac{1}{2}I + K)^{-1}f) \text{ in } \Omega,$$

*and there is a uniform estimate*

(5.75) $$\|\nabla u^*\|_{L^p(\partial\Omega)} \leq C_p \|f\|_{H^{1,p}(\partial\Omega)}.$$

REMARK. The proof of Theorem 5.10 essentially follows the proof of Theorem 5.8. The only difference is that we use appropriate results for Hardy space instead. These can be found in the previous chapter. There also is an analogue of Corollary 5.9 about invertibility of

$$S : \hbar^p(\partial\Omega) \to H^{1,p}(\partial\Omega) . \tag{5.76}$$

Thus our solution to (5.73) can be also written as

$$u = \mathcal{S}(S^{-1}f). \tag{5.77}$$

CHAPTER 6

# The Equivalence of Hardy Space Definitions

## 6.1. Preliminaries

In this chapter we present the main result of this work, the fact that the atomic definition and the definition by conjugate harmonic functions of Hardy space are equivalent even on Riemannian manifolds. The result is established for $C^1$ domains in any dimension and for Lipschitz domains in dimension 3.

First we recall known result from $\mathbb{R}^n$. For further reference see [25]. Let $(n-1)/n < p \leq 1$. Consider a harmonic function $u$ defined on the upper half-space

(6.1) $$\mathbb{R}^{n+1}_+ = \{(x,t) : x \in \mathbb{R}^n, t > 0\}.$$

Assume also that $u$ satisfies

(6.2) $$\sup_{t>0} \int_{\mathbb{R}^n} |\nabla u(x,t)|^p dx < \infty,$$

where the gradient in the formula (6.2) is considered in both variables $x$ and $t$. Then $u$ has well defined normal derivative $\partial_\nu u$ on $\partial \mathbb{R}^{n+1}_+$, this derivative belongs to $\hbar^p_{\mathrm{at}}(\mathbb{R}^n)$ and

(6.3) $$\left\|\tfrac{\partial u}{\partial \nu}\right\|^p_{\hbar^p_{\mathrm{at}}(\mathbb{R}^n)} = \left\|\tfrac{\partial}{\partial t}u\big|_{\partial\mathbb{R}^{n+1}_+}\right\|^p_{\hbar^p_{\mathrm{at}}(\mathbb{R}^n)} \approx \|(\nabla u)^*\|^p_{L^p(\mathbb{R}^n)} \approx \sup_{t>0} \int_{\mathbb{R}^n} |\nabla u(x,t)|^p dx.$$

The goal is to obtain similar claim for variable coefficient setting on a smooth compact Riemannian manifolds with a metric tensor of class $C^{1,1}$.

Take $u \in C^1_{\mathrm{loc}}(\Omega)$ to be a solution to the Laplace-Beltrami equation

(6.4) $$\Delta u = 0 \text{ in } \Omega,$$

where $\Omega$ is an open domain in $M$ with $C^1$ boundary.

We approximate $\Omega$ by a increasing sequence $\Omega_1 \subset \Omega_2 \subset \cdots \subset \Omega$ of $C^1$ domains approximating $\Omega$ from inside. Assume also that on any small neighborhood of a boundary point (2.3) and (2.4) hold. We say that such sequence of domains approximates $\Omega$ in $C^1$.

The condition analogous to (6.2) is

(6.5) $$\sup_{n \in N} \int_{\partial \Omega_n} |\nabla u(x)|^p \, d\sigma_n(x) < \infty,$$

where $d\sigma_n$ is the surface measure on $\partial \Omega_n$.

First, we will give our argument for slightly "better" domains; namely consider a continuous family $\Omega_t$, $t > 0$ of domains defined precisely in Chapter 3. That is $\Omega_t$, $u_t$ and $d\sigma_t$ are defined as in Chapter 3 by pulling back the metric tensor $g$

using a flow $\mathcal{F}_t$ generated by some smooth vector field $\Theta$ on $M$ pointing inside $\Omega$. In this setting instead of (6.5) we require

$$(6.6) \qquad \sup_{t>0} \int_{\partial \Omega_t} |\nabla u(x)|^p d\widetilde{\sigma}_t(x) = \sup_{t>0} \int_{\partial \Omega} |\nabla u_t(x)|^p \, d\sigma_t(x) < \infty.$$

The result we establish is known in $\mathbb{R}^n$ for a flat Laplacian. Dahlberg in [7] proved that for $n = 3$ the condition (6.6) with $p = 1$ implies that $\partial_\nu u \in \hbar^1(\partial\Omega)$, provided $\Omega \subset \mathbb{R}^n$ is a Lipschitz domain with connected boundary. It not known whether same thing remains true for $n \geq 4$ (c.f. problem #2 in [7]). If in addition the domain is $C^1$ the answer is positive in any dimension.

New in our proof is the idea of replacing subharmonicity of $|\nabla u|^q$ by a weaker notion of $C$-subharmonicity. As we will see later this brings several complications.

## 6.2. C-subharmonicity

Recall that for a flat Laplace operator on $\mathbb{R}^{n+1}$ and any harmonic function $u$, the function $|\nabla u|^q$ is subharmonic, i.e., for $q \geq (n-1)/n$ and $|\nabla u| > 0$ we have

$$(6.7) \qquad \Delta(|\nabla u|^q) \geq 0.$$

We would like to get similar result in our setting, i.e., for $u$ which solves $\Delta u = 0$ on $M$. It is clear that a statement like (6.7) will not work here, because in general $M$ is not flat. The solution is to define a new weaker variant of (6.7).

DEFINITION 6.1. Let $\mathcal{O} \subset M$ be an open set and $F$ a $C^1(\mathcal{O})$ function. Let $C$ be a nonnegative constant. We say that $F$ is $C$-subharmonic on $\mathcal{O}$, if for any test function $\varphi \in C_0^\infty(\mathcal{O})$, $\varphi \geq 0$ we have

$$(6.8) \qquad -\langle \nabla F, \nabla \varphi \rangle + C \langle F, \varphi \rangle \geq 0.$$

REMARK 6.2. If the metric tensor and the function $F$ are of class $C^2$ then the definition 6.1 is equivalent to the statement

$$(6.9) \qquad \Delta F + CF \geq 0.$$

Notice also that 0-subharmonicity of $F$ is just the standard subharmonicity i.e. $\Delta F \geq 0$.

From now on, we assume that the metric tensor $g$ is of class $C^{1,1}$, i.e., the gradient $\nabla g$ is Lipschitz.

Pick a point $x \in \mathcal{O}$ and take the geodesic coordinates centered at the point $x$ on some small open neighborhood of $x$. Compute $\Delta(|\nabla u|^q)$ for $u$ satisfying $\Delta u = 0$. At the point $x$ we have

$$(6.10) \qquad g_{ij}(x) = \delta_{ij}, \qquad \partial_k g_{ij}(x) = 0.$$

If the function $u$ does not have enough regularity, we will understand all derivatives in the computation bellow in the sense of distributions. Let $(g^{ij})$ be a matrix inverse to $(g_{ij})$. Denote by $X$ the gradient $\nabla u$. Using the summation convention

$$(6.11) \qquad X^i = g^{ij} \partial_j u, \qquad i = 1, 2, \ldots, n.$$

Finally, denote by $F$ the function $|\nabla u|^2$, i.e.,

$$(6.12) \qquad F = (g_{ij} X^i X^j).$$

We get:

$$\text{(6.13)} \qquad g^{lk}g^{1/2}\partial_k F^{\frac{q}{2}} = g^{lk}g^{1/2}\frac{q}{2}F^{\frac{q-2}{2}}\left\{(\partial_k g_{ij})X^i X^j + 2g_{ij}(\partial_k X^i)X^j\right\}.$$

Hence

$$\Delta(F^{\frac{q}{2}}) = g^{-1/2}\frac{q}{2}F^{\frac{q-4}{2}}\Big\{\partial_l(g^{lk}g^{1/2})F\left[(\partial_k g_{ij})X^i X^j + 2g_{ij}(\partial_k X^i)X^j\right]$$
$$+ g^{lk}g^{1/2}F\left[(\partial^2_{kl} g_{ij})X^i X^j + 2(\partial_k g_{ij})(\partial_l X^i)X^j\right.$$
$$\left. + 2g_{ij}(\partial^2_{kl}X^i)X^j + 2g_{ij}(\partial_k X^i)(\partial_l X^j)\right]$$
$$\text{(6.14)}$$
$$+ \tfrac{q-2}{2}\left[(\partial_k g_{ij})X^i X^j + 2g_{ij}(\partial_k X^i)X^j\right]\left[(\partial_l g_{mn})X^m X^n + 2g_{mn}(\partial_l X^n)X^m\right]\Big\}.$$

Now if we evaluate (6.14) at the point $x$ using (6.10) we get

$$\Delta(F^{\frac{q}{2}}) = qF^{\frac{q-4}{2}}\sum_{i,k}\left\{F\left[\tfrac{1}{2}(\partial^2_k g_{ij})X^i X^j + (\partial^2_k X^i)X^i + (\partial_k X^i)^2\right]+\right.$$
$$\text{(6.15)} \qquad\qquad\qquad \left. + (q-2)\left[(\partial_k X^i)X^i\right]^2\right\}.$$

Using the estimate from Chapter 7 paragraph 3 of [**27**] we have that

$$\text{(6.16)} \qquad \sum_{i,k}\left((q-2)\left[(\partial_k X^i)X^i\right]^2 + F(\partial_k X^i)^2\right) \geq C_q F|\nabla X|^2,$$

where slightly abusing notation $|\nabla X|^2 = \sum_{i,k}(\partial_k X^i)^2$. The constant $C_q = \frac{1}{q}(1 + (q-2)(\frac{n-1}{n})) > 0$ provided $q > (n-2)/(n-1)$.

Now we have to deal with the residual terms. Fist of all we need to use the assumption that $u$ is harmonic. Using (6.11) we get

$$\text{(6.17)} \qquad \begin{aligned}\partial_k X^i &= (\partial_k g^{im})\partial_m u + g^{im}(\partial^2_{km}u)\\ \partial^2_k X^i &= (\partial^2_k g^{im})\partial_m u + 2(\partial_k g^{im})\partial^2_{km}u + g^{im}(\partial^3_{kkm}u).\end{aligned}$$

Evaluating this at $x$ gives

$$\text{(6.18)} \qquad \partial^2_k X^i = (\partial^2_k g^{im})\partial_m u + \partial^3_{kki}u.$$

Now

$$0 = \partial_k(\Delta u) = \partial_k\left(g^{1/2}\partial_j(g^{ji}g^{1/2}\partial_i u)\right) =$$
$$\text{(6.19)}$$
$$= (\partial_k g^{1/2})\left(\partial_j(g^{ji}g^{1/2}\partial_i u)\right) + g^{1/2}\left[\partial^2_{kj}(g^{ji}g^{1/2})\partial_i u + g^{ji}g^{1/2}\partial^3_{kji}u\right].$$

Evaluating this at $x$ gives

$$\text{(6.20)} \qquad \partial^3_{ikk}u = -\partial^2_{ij}(g^{jk}g^{1/2})\partial_k u.$$

Now we plug-in (6.20) into (6.18) and then use (6.18) together with (6.16) to estimate (6.15). We get

$$\text{(6.21)} \qquad \Delta(F^{\frac{q}{2}}) \geq qF^{\frac{q-2}{2}}\left(C_q|\nabla X|^2 + A(\nabla^2 g, X, X)\right),$$

where $A$ is certain trilinear form. Notice, that eventually we got that $\Delta(F^{\frac{q}{2}})$ is actually a function, i.e., not only a distribution. This follows from the fact that on the right side of (6.21) we differentiate $u$ at most twice, i.e., if we assume $C^{1,1}$ regularity of the metric tensor, then $u$ has two classical derivatives.

Now since $\nabla g$ is Lipschitz, it follows that $A$ can be estimated by

(6.22) $$|A(\nabla^2 g, X, X)| \leq C|X|^2,$$

where $C$ is some very big constant depending on $\nabla^2 g$. Finally (6.22) together with (6.21) gives (since $F = |X|^2$):

(6.23) $$\Delta(F^{\frac{q}{2}}) + K_q F^{\frac{q}{2}} \geq (C_q q) F^{\frac{q-2}{2}} |\nabla X|^2 \geq 0.$$

Let us remark that although $A$ naturally depends on the chosen point $x \in \mathcal{O}$, it is clear that the bound (6.22) can be done uniformly for all $x$, since the function $A$ depends only on the metric tensor $g$ and its first two derivatives, hence provided $g \in C^{1,1}$ everything works. Therefore we have established:

PROPOSITION 6.3. *Let the metric tensor $g$ on Riemannian manifold $M$ be of class $C^{1,1}$. Then for any $q > (n-2)/(n-1)$ ($n = \dim M$) there is a constant $K_q > 0$ such that for any solution $u$ to the equation*

(6.24) $$\Delta u = 0 \text{ in } \mathcal{O}, \qquad \mathcal{O} \subset M, \qquad \mathcal{O} \text{ open},$$

*the function $|\nabla u|^q$ is $K_q$-subharmonic at any point $x \in \mathcal{O}$ at which $\nabla u \neq 0$, i.e.,*

(6.25) $$\Delta(|\nabla u|^q) + K_q(|\nabla u|^q) \geq 0.$$

## 6.3. The main step

We begin by defining a function $F: \Omega \to \mathbb{R}^+$ by

(6.26) $$F(x) = |\nabla u(x)|.$$

The we claim that $F \in L^p(\Omega)$, granted (6.6). Proving this is not difficult, if we integrate (6.6) in $t$ between $(0, \varepsilon]$, we get that $F \in L^p(\mathcal{C})$ where $\mathcal{C}$ is some small inside collar neighborhood of $\partial\Omega$. The $C^1$ regularity of $u$ inside $\Omega$ gives us same in $\Omega \setminus \mathcal{C}$.

Pick any $q$ such that $(n-2)/(n-1) < q < p$. We see that $F^q$ belongs a certain space $L^r(\Omega)$ with $r = p/q > 1$. On $\Omega$ we can solve the Dirichlet problem

(6.27) $$-\Delta v = F^q \text{ in } \Omega, \qquad v\big|_{\partial\Omega} = 0,$$

which gives us a solution $v \in L^r(\Omega)$. Actually $v$ has slightly better regularity that just $L^r(\Omega)$. In general, we introduce the following notation. By $\mathcal{G} f$ we denote a unique function $u$ that solves $-\Delta u = f$, $u\big|_{\partial\Omega} = 0$. Hence, $v = \mathcal{G}(F^q)$. Notice that $v \geq 0$, since $\mathcal{G}$ is a positive operator.

By Proposition 6.3 we have that

(6.28) $$G = F^q - K_q v \qquad \text{is subharmonic, i.e.,} \quad \Delta G \geq 0.$$

Here $K_q$ is the constant from Proposition 6.3 depending on $q$ and $M$.

Using the flow $\mathcal{F}_t$ mentioned above, we can parameterize the collar neighborhood of $\partial\Omega$. For any $x \in \mathcal{C}$ we can write $x = (x', t)$, where $x' \in \partial\Omega$ and $t \geq 0$ are characterized by

(6.29) $$x = \mathcal{F}_t x'.$$

Denote by $P^t$ the solution operator to the Dirichlet problem

(6.30) $$\Delta u = 0 \text{ in } \Omega_t, \qquad u\big|_{\partial \Omega_t} = f,$$

whose properties has been studied in Chapter 3. Since the function $G$ is subharmonic we claim that for each $x' \in \partial\Omega$, $t > 0$ and $\varepsilon > 0$ we have

(6.31) $$G(x', t + \varepsilon) \leq P^\varepsilon(G(., \varepsilon))(x', t).$$

Seeing this is not difficult. Clearly the both function coincides on $\partial\Omega_t$ and are continuous there. Therefore $P^\varepsilon(G(.,\varepsilon)) \in C(\overline{\Omega_t})$. This implies that both functions are bounded on $\partial\Omega \times \mathbb{R}^+$. Now a variant of the maximum principle, since the function on the left is subharmonic and the function on the right is harmonic gives (6.31).

The assumption (6.6) and the remark that follows Theorem 3.4 (the observation that $v\big|_{\partial\Omega_t}$ are uniformly bounded in the $L^r$ norm) give us that the functions $G(.,\varepsilon)$ are uniformly bounded in $L^r(\partial\Omega)$, for some $r > 1$. Reflexivity of the space $L^r$ therefore implies that we can find a function $h \in L^r(\partial\Omega)$ such that $G(.,\varepsilon_k) \to h$ weakly in $L^r$, as $k \to \infty$. As we saw in the proof of Theorem 3.1 this allows us to pass to the limit $k \to \infty$ on the right hand side of (6.31) and gives

(6.32) $$G(x', t) \leq P^0(h)(x', t) \qquad \forall t > 0.$$

Notice also $h \geq 0$ and

(6.33) $$\|h\|_{L^r(\partial\Omega)}^r \leq C \sup_{t>0} \int_{\partial\Omega_t} |\nabla u(x)|^p d\widetilde{\sigma}_t(x),$$

since the norm of both functions defining $G$ can be estimated by the right side of (6.33). Another way to write (6.32) is

(6.34) $$|\nabla u|^q \leq P^0(h) + K_q v = P^0(h) + K_q \mathcal{G}(|\nabla u|^q).$$

Notice, that in the flat case (i.e., when $K_q = 0$) the estimate (6.34) is enough to bound the maximal operator of $|\nabla u|^q$, since the bound on the maximal operator of $P^0(h)$ is known from Chapter 3.

If $K_q > 0$ this is no longer true, because we do not have bound on the maximal operator of $\mathcal{G}(|\nabla u|^q)$. But a nice thing about (6.34) is that it can be iterated, since $\mathcal{G}$ is a positive operator. For example iterating (6.34) once gives

(6.35) $$|\nabla u|^q \leq P^0(h) + K_q \mathcal{G}(P^0(h)) + K_q^2 \mathcal{G}^2(|\nabla u|^q).$$

Here $\mathcal{G}^2 = \mathcal{G} \circ \mathcal{G}$. In general, the $(n-1)$-th iteration of (6.34) looks as follows:

(6.36) $$\begin{aligned}|\nabla u|^q \leq &P^0(h) + K_q \mathcal{G}(P^0(h)) + K_q^2 \mathcal{G}^2(P^0(h)) + \cdots + \\ &+ K_q^{n-1}\mathcal{G}^{n-1}(P^0(h)) + K_q^n \mathcal{G}^n(|\nabla u|^q).\end{aligned}$$

The key is to realize that starting with $|\nabla u|^q \in L^r(\Omega)$ for some $r > 1$, the operator $\mathcal{G}$ smoothness things out, that is, there is an integer $n$ for which $\mathcal{G}^n(|\nabla u|^q)$ is a continuous function on $\overline{\Omega}$. Naturally, the maximal operator of continuous function on $\overline{\Omega}$ can be estimated trivially. On the other hand, (3.3) gives us estimate on the maximal operator of the harmonic function $P^0(h)$. By invoking Theorem B.10 we get an estimate on the maximal operator of $\mathcal{G}^i(P^0(h))$ for $i = 1, 2, \ldots, n-1$. This gives

$$\text{(6.37)} \quad (\nabla u)^* \leq C\big[((P^0(h))^*)^{1/q} + (\mathcal{G}(P^0(h))^*)^{1/q} + \cdots + \\ + (\mathcal{G}^{n-1}(P^0(h))^*)^{1/q} + (\mathcal{G}^n(|\nabla u|^q)^*)^{1/q}\big],$$

and therefore

$$\text{(6.38)} \quad \int_{\partial\Omega} ((\nabla u)^*)^p \, d\sigma \leq C(\|h\|^r_{L^r(\partial\Omega)} + \|F^q\|_{L^r(\Omega)}) \leq C \sup_{t>0} \int_{\partial\Omega_t} |\nabla u(x)|^p d\widetilde{\sigma}_t(x).$$

We briefly comment on the equivalence of conditions (6.5) and (6.6). The crutial point in the argument above is that the constant $C_p$ in Theorems 3.1 and 3.3 is independent of $t$, i.e., the estimates (3.3) and (3.18) remains valid for any $\Omega_t$. This follows from Lemma 3.2. We claim that same can be said about the domains $\Omega_1 \subset \Omega_2 \subset \ldots$, provided they are chosen as described at the beginning of this chapter. The proof uses the fact that there is $k \in N$ such that for $n \geq k$ we have 1-1 correspondence between the points on $\partial\Omega_n$ and $\partial\Omega$ via the flow $\mathcal{F}_t$. Hence, we can compare the measures $d\sigma_n$ and $d\sigma$ and get that the Radon-Nikodym derivative $\rho_n = \frac{d\sigma_n}{d\sigma}$ is a continuous function converging to 1 in $C(\partial\Omega)$ norm as $n \to \infty$. This is enough to establish a lemma analogous to Lemma 3.2 for the domains $\Omega_n$ (for $p > 1$). Having this, the other key element of our argument was the limiting process (6.31) for $\varepsilon \to 0$. We can repeat the same for the domains $\Omega_n$ (only the notation will be a bit more complicated). Thus, same conclusion follows even for condition (6.5). The final comment is that the numbers (6.5) and (6.6) are comparable. Really, take any $\Omega_t \subset \Omega_n$ and use (4.2) when $p > 1$ and (4.6) when $(n-1)/n < p \leq 1$ to get for such $\Omega_t \subset \Omega_n$

$$\text{(6.39)} \quad \int_{\partial\Omega_t} |\nabla u(x)|^p d\widetilde{\sigma}_t(x) \leq \int_{\partial\Omega_t} |(\nabla u)^*(x)|^p d\widetilde{\sigma}_t(x) \approx \\ \approx \|(\nabla u)^*\|_{L^p(\Omega_n)} \leq \|(\nabla u)^*\|_{L^p(\Omega)} \leq C \sup_{n\in N} \int_{\partial\Omega_n} |\nabla u(x)|^p \, d\sigma_n(x).$$

And vice versa for different domains where $\Omega_n \subset \Omega_t$. This establishes

PROPOSITION 6.4. *Let* $(n-1)/n < p < \infty$ *and let* $\Omega_1 \subset \Omega_2 \subset \cdots \subset \Omega$ *be an infinite sequence of* $C^1$ *domains increasing to* $\Omega$ *that approximates* $\Omega$ *in* $C^1$. *Suppose that a function* $u \in C^1_{\text{loc}}(\Omega)$ *is harmonic, i.e., it solves* $\Delta u = 0$ *on* $\Omega$ *and satisfies*

$$\text{(6.40)} \quad \sup_{n\in N} \int_{\partial\Omega_n} |\nabla u(x)|^p \, d\sigma_n(x) < \infty.$$

*Then such function has well-defined values of* $\nabla u$ *on* $\partial\Omega$ *and moreover*

$$\text{(6.41)} \quad (\nabla u)^* \in L^p(\partial\Omega) \quad \text{and} \quad \|(\nabla u)^*\|^p_{L^p(\partial\Omega)} \approx \sup_{n\in N} \int_{\partial\Omega_n} |\nabla u(x)|^p \, d\sigma_n(x).$$

PROOF. Everything else except inequality $\geq$ in (6.41) has already been established. However, since this inequality hold trivially, the Proposition follows. $\square$

## 6.4. The equivalence theorem on $C^1$ domains

If $p > 1$ we can immediately claim:

THEOREM 6.5. *Let $1 < p < \infty$ and let the metric tensor on $M$ be of class $C^{1,1}$. Assume also that $u \in C^1_{\text{loc}}(\Omega)$ is harmonic in $\Omega \subset M$, where $\Omega$ is an open set with $C^1$ boundary.*

*The function $g = \frac{\partial u}{\partial \nu}$ belongs to $L^p(\partial\Omega)$ if and only if*

$$(6.42) \qquad \sup_{n \in N} \int_{\partial\Omega_n} |\nabla u(x)|^p \, d\sigma_n(x) < \infty,$$

*for a series of $C^1$ domains $\Omega_1 \subset \Omega_2 \subset \Omega_3 \subset \ldots$ increasing to $\Omega$ that approximates $\Omega$ in $C^1$. Moreover,*

$$(6.43) \qquad \|g\|^p_{L^p(\partial\Omega)} \approx \sup_{n \in N} \int_{\partial\Omega_n} |\nabla u(x)|^p \, d\sigma_n(x).$$

PROOF. The "if" part of the theorem follows from the uniqueness result for the Neumann boundary problem established in Theorem 5.5 of [**23**] in conjunction with Theorem 4.1 of this paper. The "only if" part follows from Proposition 6.4. □

Before we produce a result analogous to Theorem 6.5 for $(n-1)/n < p \leq 1$ we need to prove the following.

PROPOSITION 6.6. *Let $(n-1)/n < p \leq 1$ and let the metric tensor be of class $C^\alpha$ for some $\alpha > 0$. Assume also that $u$ is a harmonic function in $\Omega \subset M$ where $\Omega$ is an open set with $C^1$ boundary. Then $(\nabla u)^*$ belongs to $L^p(\partial\Omega)$ if and only if the function $g = \frac{\partial u}{\partial \nu}$ belongs to $\hbar^p(\partial\Omega)$ and $u$ can be written as*

$$(6.44) \qquad u = \mathcal{S}((-\tfrac{1}{2}I + K^*)g).$$

*The equality (6.44) is understood modulo constants.*

PROOF. By Theorem 4.2 if the function $g$ belongs to $\hbar^p(\partial\Omega)$ then $u$ of the form (6.44) solves $\Delta u = 0$ and also $(\nabla u)^* \in L^p(\partial\Omega)$. This does the "only if" part. The "if" part of the proposition requires to show that given $(\nabla u)^* \in L^p(\partial\Omega)$ we have $g = \frac{\partial u}{\partial \nu} \in \hbar^p(\partial\Omega)$. Once having this define we $v = \mathcal{S}((-\tfrac{1}{2}I + K^*)g)$. The uniqueness result in Theorem 4.2 implies that $u = v$ modulo constants.

The argument that $g = \frac{\partial u}{\partial \nu} \in \hbar^p(\partial\Omega)$ uses again the approach developed in [**31**]. Given a harmonic function with $(\nabla u)^* \in L^p(\partial\Omega)$, the function $g = \frac{\partial u}{\partial \nu}$ is in $\hbar^p_{\text{at}}(\partial\Omega)$, since it can be decomposed into $p$-atoms. □

Propositions 6.4 and 6.6 finally yield the main result of this chapter for $C^1$ domains.

THEOREM 6.26. *Let $(n-1)/n < p \leq 1$ and let the metric tensor on $M$ be of class $C^{1,1}$. Assume also that $u \in C^1_{\text{loc}}(\Omega)$ is harmonic, i.e., it solves $\Delta u = 0$ on $\Omega \subset M$, where $\Omega$ is an open set with $C^1$ boundary. If*

$$(6.45) \qquad \sup_{n \in N} \int_{\partial\Omega_n} |\nabla u(x)|^p \, d\sigma_n(x) < \infty,$$

*for an series of $C^1$ domains $\Omega_1 \subset \Omega_2 \subset \Omega_3 \subset \ldots$ increasing to $\Omega$ that approximates $\Omega$ in $C^1$, then $g = \frac{\partial u}{\partial \nu}$ belongs to $\hbar^p(\partial\Omega)$; the function $u$ can be written as $u =$*

$\mathcal{S}((-\frac{1}{2}I + K^*)g)$ (modulo constants) and the $\hbar^p(\partial\Omega)$ norm of $g$ is comparable with (6.45), i.e.,

$$\text{(6.46)} \qquad \|g\|^p_{\hbar^p(\partial\Omega)} \approx \sup_{n \in N} \int_{\partial\Omega_n} |\nabla u(x)|^p \, d\sigma_n(x).$$

Conversely, if $g \in \hbar^p_{\text{at}}(\partial\Omega)$ then the harmonic function $u = \mathcal{S}((-\frac{1}{2}I + K^*)g)$ satisfies (6.45) for any series of $C^1$ domains $\Omega_1 \subset \Omega_2 \subset \Omega_3 \subset \ldots$ increasing to $\Omega$ and approximating $\Omega$ in $C^1$. Also

$$\text{(6.47)} \qquad \|g\|^p_{\hbar^p(\partial\Omega)} \approx \|(\nabla u)^*\|^p_{L^p(\partial\Omega)} \approx \sup_{n \in N} \int_{\partial\Omega_n} |\nabla u(x)|^p \, d\sigma_n(x).$$

## 6.5. The equivalence theorem on Lipschitz domains

Finally, we would like to consider the case when the boundary $\partial\Omega$ is only Lipschitz. As follows from the results in [22], [23] and [24], on Lipschitz domains we can solve the Dirichlet problem on $\partial\Omega$ for initial data $f \in L^p(\partial\Omega)$ with $2 - \varepsilon < p < \infty$. If we consider the Neumann problem, same is true for initial data $g \in L^p(\partial\Omega)$ for $1 < p < 2 + \varepsilon$ and $g \in \hbar^p(\partial\Omega)$ for $1 - \varepsilon < p \leq 1$.

A second look at the proof above reveals that we had to solve the Dirichlet problem for Laplace equation with $L^r(\partial\Omega)$ boundary conditions. Here the number $r$ is given as $r = p/q$, where $(n-2)/(n-1) < q < p$. Thus if this $r$ is sufficiently close to 2, the proof above would work even for Lipschitz domains. For Hardy spaces, this happens when $\dim M = 3$. For simplicity we formulate our results only for domains $\Omega_t$.

THEOREM 6.27. *Let the metric tensor on $M$ be of class $C^{1,1}$. Consider any $1 < p < \infty$. Let $u \in C^1_{\text{loc}}(\Omega)$ be harmonic on $\Omega \subset M$, where $\Omega$ is an open set with Lipschitz boundary. If*

$$\text{(6.48)} \qquad \sup_{t > 0} \int_{\partial\Omega_t} |\nabla u(x)|^p \, d\widetilde{\sigma}_t(x) < \infty,$$

*then the function $g = \frac{\partial u}{\partial \nu}$ belongs to $L^p(\partial\Omega)$ and*

$$\text{(6.49)} \qquad \|g\|^p_{L^p(\partial\Omega)} \approx \sup_{t > 0} \int_{\partial\Omega_t} |\nabla u(x)|^p \, d\widetilde{\sigma}_t(x).$$

PROOF. There are two cases: If $2 \leq p < \infty$ the technique described above based on subharmonic majorization and the solvability of $L^p$ Dirichlet problem works. If $1 < p \leq 2$ the idea is, that granted (6.31), the natural estimate accompanying the $H^{1,p}$ regularity problem for Laplacian used on each approximating domain $\Omega_t$ yields

$$\text{(6.50)} \qquad \|(\nabla u|_{\Omega_t})^*\|_{L^p(\partial\Omega_t)} \leq C,$$

with $C$ independent of $t > 0$. (See [23]). This naturally, yields $\|(\nabla u)^*\|_{L^p(\partial\Omega)} \leq C < \infty$, as desired. The argument for $1 < p \leq 2$ is due to Marius Mitrea, to whom I am very grateful for pointing it out. □

As indicated, when $\dim M = 3$ we also have:

THEOREM 6.28. *Let dim $M = 3$ and let the metric tensor on $M$ be of class $C^{1,1}$. Assume also that $u$ is a harmonic function in $\Omega$, where $\Omega$ is an open set with Lipschitz boundary. Then there exists $\varepsilon = \varepsilon(\partial\Omega) > 0$ such that for $1 - \varepsilon < p \leq 1$ we have the following. If*

$$\sup_{t>0} \int_{\partial\Omega_t} |\nabla u(x)|^p \, d\tilde{\sigma}_t(x) < \infty, \tag{6.51}$$

*then $g = \frac{\partial u}{\partial \nu}$ belongs to $\hbar^p(\partial\Omega)$; the function $u$ can be written as $u = \mathcal{S}((-\frac{1}{2}I + K^*)g)$ (modulo constants) and the $\hbar^p(\partial\Omega)$ norm of $g$ is comparable with (6.32), i.e.,*

$$\|g\|_{\hbar^p(\partial\Omega)}^p \approx \sup_{t>0} \int_{\partial\Omega_t} |\nabla u(x)|^p \, d\tilde{\sigma}_t(x). \tag{6.52}$$

*Conversely, given $g \in \hbar_{at}^p(\partial\Omega)$ the harmonic function $u = \mathcal{S}((-\frac{1}{2}I + K^*)g)$ satisfies (6.32). Also*

$$\|g\|_{\hbar^p(\partial\Omega)}^p \approx \|(\nabla u)^*\|_{L^p(\partial\Omega)}^p \approx \sup_{t>0} \int_{\partial\Omega_t} |\nabla u(x)|^p \, d\tilde{\sigma}_t(x). \tag{6.53}$$

Whether the claim holds for dim $M \geq 4$ remains to be seen. We suspect, that the restriction on dimension of $M$ in Theorem 6.28 is not necessary. However, our proof obviously such restriction requires.

# APPENDIX A

# Variable Coefficient Cauchy Integrals

In this appendix we would like to establish certain results about Cauchy integrals on Lipschitz surfaces. These results are used throughout this work. Recall first Theorem 1.1 of [**22**] which was obtained using results of Coifman, McIntosh and Meyer [**5**].

THEOREM A.1. *Let $\Gamma$ be a Lipschitz graph in $\mathbb{R}^n$ of the form $x_n = \phi(x_1, \ldots, x_n)$ for some Lipschitz function $\phi : \mathbb{R}^{n-1} \to \mathbb{R}$. There exists $N = N(n)$ such that if $k \in C^N(\mathbb{R}^n \setminus 0)$ is an odd function ($k(-x) = -k(x)$) and homogeneous of degree $-(n-1)$, then $k(x-y)$ is a kernel of an operator $K$ bounded on $L^p(\Gamma)$ for $1 < p < \infty$, of norm*

$$\text{(A.1)} \qquad \|K\|_{\mathcal{L}(L^p)} \leq C(p, \Gamma) \|k\|_{S^{n-1}} \|_{C^N}.$$

Here $L^p(\Gamma)$ is defined using the surface measure (i.e. $(n-1)$ dimensional Hausdorff measure) on $\Gamma$ and the constant $C(p, \Gamma)$ depends only on $p$ and $\|\nabla\phi\|_{L^\infty}$.

If we write the operator (A.1) in coordinates we obtain

$$\text{(A.2)} \qquad Kf(x) = \text{P.V.} \int_{\mathbb{R}^{n-1}} k(x-y, \phi(x) - \phi(y)) f(y) \sqrt{1 + |\nabla\phi|^2} \, dy.$$

Now want specifically emphasize the dependence of (A.2) on the function $\phi$, namely write $K^\phi$ instead of just $K$. The main goal of this appendix is to establish that

$$\text{(A.3)} \qquad \phi \mapsto K^\phi$$

is a continuous map from $Lip(\mathbb{R}^{n-1})$ to $\mathcal{L}(L^p)$. Here $Lip(\mathbb{R}^{n-1})$ means a Banach space of Lipschitz functions with the norm $\|\phi\|_{L^\infty} + \|\nabla\phi\|_{L^\infty}$. We want to show that

$$\text{(A.4)} \qquad \|K^{\phi_1} - K^{\phi_2}\|_{\mathcal{L}(L^p)} \leq C(\|\nabla\phi_1\|_{L^\infty}) \|\phi_1 - \phi_2\|_{Lip(\mathbb{R}^{n-1})}.$$

We use the work [**5**] and the method of rotations to establish (A.4).

THEOREM A.2. *Let $\Gamma$ be a Lipschitz graph in $\mathbb{R}^n$, as in Theorem A.1. Let $A : \mathbb{R}^n \to \mathbb{R}$ be another Lipschitz function. There exists $N = N(n)$ such that, if $k \in C^N(\mathbb{R}^n \setminus 0)$ is even, i.e., $k(-x) = k(x)$ and homogeneous of degree $-n$, then*

$$\text{(A.5)} \qquad K(x, y) = k(x - y)(A(x) - A(y))$$

*is a kernel of an operator bounded on $L^p(\Gamma)$ for $1 < p < \infty$, of norm*

$$\text{(A.6)} \qquad \|K\|_{\mathcal{L}(L^p)} \leq C(p, \Gamma) \|\nabla A\|_{L^\infty} \|k\|_{S^{n-1}} \|_{C^N}.$$

Here $L^p(\Gamma)$ as before is defined using surface measure. Again the constant $C(p, \Gamma)$ in fact depends only on $p$ and $\|\nabla\phi\|_{L^\infty}$.

Now, our result follows from Theorem A.2.

PROPOSITION A.3. *There exists $M = M(n)$ such that if $k \in C^M(\mathbb{R}^n \setminus 0)$, then the map (A.3) is continuous, i.e., (A.4) holds.*

PROOF. Pick any $\phi_1, \phi_2$ from $Lip(\mathbb{R}^{n-1})$. We want to estimate the difference $K^{\phi_1} - K^{\phi_2}$. We do it in several steps. First define $\psi^\lambda(x) = (1-\lambda)\phi_1 + \lambda\phi_2$ for $0 \leq \lambda \leq 1$. Look first at the operator

$$(A.7) \quad B^\lambda f(x) = \text{P.V.} \int_{\mathbb{R}^{n-1}} k(x-y, \psi^\lambda(x) - \psi^\lambda(y)) f(y) \sqrt{1+|\nabla \phi_1|^2} \, dy.$$

Clearly, by Theorem A.1, $B^\lambda$ is a family of uniformly bounded operators. Formally,

$$(A.8) \quad \frac{d}{d\lambda} B^\lambda f(x) = \text{P.V.} \int_{\mathbb{R}^{n-1}} \frac{\partial}{\partial_n} k(x-y, \psi^\lambda(x) - \psi^\lambda(y)) \times \\ ((\phi_2 - \phi_1)(x) - (\phi_2 - \phi_1)(y)) f(y) \sqrt{1+|\nabla \phi_1|^2} \, dy.$$

Hence

$$(A.9) \quad \frac{d}{d\lambda} B^\lambda f(x) = \text{P.V.} \int_{\Gamma^\lambda} \frac{\partial}{\partial_n} k(x-y, \psi^\lambda(x) - \psi^\lambda(y)) \times \\ ((\widetilde{\phi_2} - \widetilde{\phi_1})(x) - (\widetilde{\phi_2} - \widetilde{\phi_1})(y)) f(y) \frac{\sqrt{1+|\nabla \phi_1|^2}}{\sqrt{1+|\nabla \psi^\lambda|^2}} \, d\sigma^\lambda(y),$$

where $\Gamma^\lambda$ is the Lipschitz surface given by graph of $\psi^\lambda$ and $\frac{\partial}{\partial_n} k(z, z_n)$ is partial derivative with respect to the last variable. The functions $\widetilde{\phi}_i$ for $i = 1, 2$ are defined by $\widetilde{\phi}_i(x) = \phi_i(x')$, where $x = (x', x_n) \in \mathbb{R}^n$, i.e, $x'$ are the first $n-1$ coordinates of $x$.

Now since $k$ is odd and homogeneous of degree $-(n-1)$ the function $\frac{\partial}{\partial_n} k$ must be even and homogeneous of degree $-n$. Thus Theorem A.2. applies and gives us

$$\|\tfrac{d}{d\lambda} B^\lambda f\|_{L^p(\Gamma^\lambda)} \leq C(p, \Gamma^\lambda) \|\nabla(\phi_2 - \phi_1)\|_{L^\infty} \|k|_{S^{n-1}}\|_{C^M} \left\| f \frac{\sqrt{1+|\nabla \phi_1|^2}}{\sqrt{1+|\nabla \psi^\lambda|^2}} \right\|_{L^p(\Gamma^\lambda)} \leq$$

$$(A.10) \quad \leq C \|\nabla(\phi_2 - \phi_1)\|_{L^\infty} \|f\|_{L^p(\Gamma^\lambda)},$$

where $C$ is independent of $\lambda$ and depends only on $\phi_1$ and $\phi_2$. Therefore we get

$$(A.11) \quad \|B^1 - B^0\|_{\mathcal{L}^p(\mathbb{R}^{n-1})} \leq C \|\nabla(\phi_2 - \phi_1)\|_{L^\infty}.$$

Now we can evaluate the difference $K^{\phi_1} - K^{\phi_2}$.

$$(A.12)$$
$$\|(K^{\phi_1} - K^{\phi_2}) f\|_{L^p} \leq \|(B^0 - B^1) f\|_{L^p} \\ + \left\| \text{P.V.} \int_{\mathbb{R}^{n-1}} k(x-y, \phi_2(x) - \phi_2\lambda(y)) f(y) (\sqrt{1+|\nabla \phi_1|^2} - \sqrt{1+|\nabla \phi_2|^2}) \, dy \right\|_{L^p}$$
$$\leq C \|\nabla(\phi_2 - \phi_1)\|_{L^\infty} \|f\|_{L^p} + C \|f(\sqrt{1+|\nabla \phi_1|^2} - \sqrt{1+|\nabla \phi_2|^2})\|_{L^p}$$
$$\leq C \|\nabla(\phi_2 - \phi_1)\|_{L^\infty} \|f\|_{L^p}.$$

This finishes our proof. □

As in [**22**], we can establish a variable coefficients variant of Proposition A.3. The following Proposition corresponds to Proposition 1.2 of [**22**].

PROPOSITION A.4. *There is $M = M(n)$ such that the following holds. Let $b(x, z)$ be odd in $z$, homogeneous of degree $-(n-1)$ in $z$, and assume $D_z^\alpha b(x, z)$ is continuous and bounded on $\mathbb{R}^n \times S^{n-1}$ for $|\alpha| \leq M$. Then $b(x, x-y)$ is a kernel of an operator $B^\phi$, bounded on $L^p(\Gamma)$ for $1 < p < \infty$ where $\Gamma$ is a graph of a Lipschitz function $\phi : \mathbb{R}^{n-1} \to \mathbb{R}$. Moreover, the map $\phi \mapsto B^\phi$, if we look at $B^\phi$ as an operator on $L^p(\mathbb{R}^{n-1})$, is a continuous function from $Lip(\mathbb{R}^{n-1})$ to $\mathcal{L}(L^p(\mathbb{R}^{n-1}))$.*

PROOF. There is no difference between our proof and the result in [**22**]. Using classical spherical decomposition of $b(x, z)$ we write

$$(A.13) \qquad b(x,z) = \sum_{j \geq 1} b_j(x) \varphi_j(z/|z|) |z|^{-(n-1)},$$

where we pick $M$ big enough so that Proposition A.3. applies and

$$(A.14) \qquad \|b_j\|_{L^\infty} \|\varphi_j\|_{C^N} \leq C j^{-2}.$$

If $k_j(z) = \varphi_j(z/|z|)|z|^{-(n-1)}$ with $\varphi_j$ odd, then by Proposition A.3 each operator $K_j^\phi$ is well defined on $L^p(\mathbb{R}^{n-1})$ and continuous in $\phi$. This gives

$$(A.15) \qquad B^\phi f(x') = \sum_{j \geq 1} b_j(x', \phi(x')) K_j^\phi(x'),$$

where we use notation $x = (x', x_n)$ with $x' \in \mathbb{R}^{n-1}$. Using continuity of $b_j$ in $x$ we get for $\phi_1, \phi_2 \in Lip(\mathbb{R}^{n-1})$:

$$(A.16) \quad |b_j(x', \phi_1(x')) - b_j(x', \phi_2(x'))| \leq \omega_j(|\phi_1(x') - \phi_2(x')|) \leq \omega_j(\|\phi_1 - \phi_2\|_{L^\infty}).$$

Here $\omega_j$ is modulus of continuity of $b_j$. (A.16) together with (A.14) allow us to estimate the difference $B^{\phi_1} - B^{\phi_2}$ and yield desired continuity. □

We also have an analogue of Proposition 1.3. of [**22**].

PROPOSITION A.5. *Under the hypothesis of Proposition A.4, $b(y, x-y)$ is a kernel of an operator $\widetilde{B}$ on $L^p(\Gamma)$ for $1 < p < \infty$. Again, if we view $\widetilde{B^\phi}$ as an operator on $L^p(\mathbb{R}^{n-1})$, we get continuity of map $\phi \to \widetilde{B^\phi}$ for $\phi \in Lip(\mathbb{R}^n)$.*

Now we briefly look at the case of Hardy spaces. We want to adapt proofs from appending B of [**23**]. First, we prove an analogue of Proposition B.1 from [**23**].

PROPOSITION A.6. *Assume as above that $\Gamma$ is a graph in $\mathbb{R}^n$ given by a Lipschitz function $\phi$. There exists $N = N(n)$ such that if $k \in C^N(\mathbb{R}^n \setminus 0)$ is odd and homogeneous of degree $-(n-1)$, then*

$$(A.17) \qquad \mathcal{K}^\phi f(x) = \int_\Gamma k(x-y) f(y) \, d\sigma(y), \qquad x \in \mathbb{R}^n \setminus \Gamma$$

*satisfies the nontangential maximal function estimate*

$$(A.18) \qquad \|(\mathcal{K}^\phi f)^*\|_{L^p(\Gamma)} \leq C(p, \Gamma) \|k|_{S^{n-1}}\|_{C^N} \|f\|_{h^p_{at}(\Gamma)},$$

for $(n-1)/n < p \leq 1$. Moreover, if $\phi_1$, $\phi_2$ are two Lipschitz functions with graphs $\Gamma_1$, $\Gamma_2$, respectively, we also have
(A.19)
$$\|(\mathcal{K}^{\phi_1}(f\rho) - \mathcal{K}^{\phi_2}f)^*\|_{L^p(\Gamma_2)} \leq C(p,\Gamma_2)\|k|_{S^{n-1}}\|_{C^N}\|\nabla(\phi_1 - \phi_2)\|_{L^\infty(\mathbb{R}^{n-1})}\|f\|_{h^p_{at}(\Gamma_2)}.$$

The function $\rho$ here is the Radon-Nikodym derivative $\rho = \frac{d\sigma_2}{d\sigma_1}$ of surface measures $d\sigma_1$, $d\sigma_2$ on $\Gamma_1$, $\Gamma_2$, respectively.

PROOF. Before we begin the proof we explain what exactly (A.19) means, since there is some ambiguity. Fix a point $x_0 \in \Gamma_2$ and consider a nontangential approach region $\gamma(x_0)$. There is exactly one point $x'_0 \in \Gamma_1$ such that $x_0$ and $x'_0$ have same first $(n-1)$ coordinates. We want to compare the maximal operator at $x_0$ with maximal operator at $x'_0$ and we do it by shifting vertically the whole domain of $\mathcal{K}^{\phi_1}(f\rho)$ so that $x_0$ and $x'_0$ coincide. By doing this, we achieve that $\gamma(x_0) = \gamma(x'_0)$ and hence we can compute $(\mathcal{K}^{\phi_1}(f\rho) - \mathcal{K}^{\phi_2}f)^*$ at $x_0$. We do this for any $x_0 \in \Gamma_2$. Naturally, in general the considered shift is different for different $x_0$. In a special case, when $\phi_1 = \phi_2 + c$ for some constant $c$, we get that mentioned shift is always by $c$ and naturally in such case we get $(\mathcal{K}^{\phi_1}(f\rho) - \mathcal{K}^{\phi_2}f)^* = 0$, which shows that (A.19) works fine is this special case. Keeping the explanation above in mind, we are ready to begin.

As in [**23**] consider normalized $p$-atoms, i.e., $f \in L^\infty(\Gamma_2)$ satisfying
(A.20)
$$\operatorname{supp} f \subset B_1(0) \cap \Gamma_2, \qquad \|f\|_{L^\infty(\Gamma_2)} \leq 1, \qquad \int_{\Gamma_2} f \, d\sigma_2 = 0,$$

and $0 \in \Gamma_2$. Since (A.18) has been established in [**23**] we concentrate on (A.19). Clearly we can view the atom (A.20) also as an atom on a another Lipschitz curve $\Gamma_1$ (although the moment condition $\int_{\Gamma_1} f \, d\sigma_1 = 0$ will not be exactly satisfied there, but it can be fixed by multiplying $f$ by a function $\rho$ (Radon-Nikodym derivative $\frac{d\sigma_2}{d\sigma_1}$) which in this case is $L^\infty$ and $\|\rho\|_{L^\infty} \approx 1$). Also $\|f\rho\|_{h^p_{at}(\Gamma_1)} \approx \|f\|_{h^p_{at}(\Gamma_2)}$.

Clearly for any $\varepsilon > 0$ there is $R$ big such that
(A.21)
$$|x| \geq R \Longrightarrow |\mathcal{K}^{\phi_1}(f\rho)(x)| \leq \varepsilon|x|^{-n},$$
$$|\mathcal{K}^{\phi_2}f(x)| \leq \varepsilon|x|^{-n}.$$

On $B_R(0)$ the $L^2$ theory due to [**5**] gives
(A.22)
$$\int_{\Gamma_2 \cap B_R(0)} [(\mathcal{K}^{\phi_1}(f\rho) - \mathcal{K}^{\phi_2}f)^*]^p \, d\sigma_2 \leq C_p \omega(\|\nabla(\phi_1 - \phi_2)\|_{L^\infty(\mathbb{R}^{n-1})}).$$

Here, we consider nontangential regions $\gamma(x)$ approaching the curve $\Gamma_2$ from above.

Combining (A.21), (A.22) yields (A.19) for atoms. Density argument brings then (A.19) for any $f \in h^p(\Gamma_2)$. □

Granted Proposition A.6 we can again establish variable coefficients extensions, the argument is same as in Proposition A.4 of [**23**].

PROPOSITION A.7. *There is $M = M(n)$ such that the following holds. Let $b(x,z)$ be odd in $z$ and homogeneous of degree $-(n-1)$ in $z$ and assume that $D_z^\alpha b(x,z)$ is continuous and bounded on $\mathbb{R}^n \times S^{n-1}$ for $|\alpha| \leq M$. Then*
(A.23)
$$\mathcal{B}f(x) = \int_\Gamma b(x, x-y) f(y) \, d\sigma(y), \qquad x \in \mathbb{R}^n \setminus \Gamma,$$

*satisfies*

(A.24) $$\|(\mathcal{B}f)^*\|_{L^p(\Gamma)} \leq C(\Gamma) \sup_{|\alpha| \leq M} \|D_z^\alpha b(x,z)\|_{L^\infty(\mathbb{R}^n \times S^{n-1})} \|f\|_{h^p(\Gamma)},$$

for $(n-1)/n < p \leq 1$. Moreover, we also have for any two Lipschitz functions $\phi_1$, $\phi_2$ with graphs $\Gamma_1, \Gamma_2$, respectively

(A.25)
$$\|(\mathcal{B}^{\phi_1}(f\rho) - \mathcal{B}^{\phi_2}f)^*\|_{L^p(\Gamma_2)} \leq$$
$$\leq C(\Gamma_2) \sup_{|\alpha| \leq M} \|D_z^\alpha b(x,z)\|_{L^\infty(\mathbb{R}^n \times S^{n-1})} \omega(\|\phi_1 - \phi_2\|_{Lip(\mathbb{R}^{n-1})}) \|f\|_{h^p_{at}(\Gamma_2)},$$

where $\rho$ is as in Proposition A.6, and $\omega$ is a modulus of continuity, i.e., a decreasing function continuous at 0 with $\omega(0) = 0$. (A.25) is understood in the sense explained at the beginning of the proof of Proposition A.6.

PROPOSITION A.8. *In Proposition A.7, assume in addition that $D_z^\alpha b(y,z)$ is Hölder continuous on $\mathbb{R}^n \times S^{n-1}$, of exponent $r > (n-1)(p^{-1} - 1)$. Let $\Gamma_0 \subset \Gamma$ is compact. Then*

(A.26) $$\widetilde{\mathcal{B}}f(x) = \int_\Gamma b(y, x-y) f(y) \, d\sigma(y), \qquad x \in \mathbb{R}^n \setminus \Gamma,$$

*satisfies*

(A.27) $$\|(\widetilde{\mathcal{B}}f)^*\|_{L^p(\Gamma_0)} \leq C \sup_{\substack{|\alpha| \leq M \\ |z|=1}} \|D_z^\alpha b(y,z)\|_{C^r(\Gamma_0)} \|f\|_{h^p(\Gamma_0)},$$

*for $(n-1)/n < p \leq 1$ and $f$ supported on $\Gamma_0$.*

*Moreover, we also have for any two Lipschitz functions $\phi_1$, $\phi_2$ with graphs $\Gamma_1$, $\Gamma_2$ respectively, an estimate similar to (A.25) on any compact $\Gamma_0 \subset \Gamma_2$, i.e.,*

(A.28)
$$\|(\widetilde{\mathcal{B}}^{\phi_1}(f\rho) - \widetilde{\mathcal{B}}^{\phi_2}f)^*\|_{L^p(\Gamma_0)} \leq C \sup_{\substack{|\alpha| \leq M \\ |z|=1}} \|D_z^\alpha b(y,z)\|_{C^r} \omega(\|\phi_1 - \phi_2\|_{Lip(\mathbb{R}^{n-1})}) \|f\|_{h^p_{at}(\Gamma_0)},$$

*where $\omega$ and $\rho$ is as in Proposition A.7 and (A.28) is understood in the sense explained at the beginning of the proof of Proposition A.6.*

Now, we prove a Proposition that will be essential in our arguments.

PROPOSITION A.9. *There is $N = N(n)$ such that the following holds. Let $g \in L^\infty$ and let $k(.)$ be an even function, homogeneous of degree $-n$, and $k \in C^N(\mathbb{R}^n \setminus \{0\})$. Then the maximal operator*

(A.29) $$K^* f(x) = \sup_{\varepsilon > 0} \left| \int_{r(x,y) > \varepsilon} k(x-y) g(y) \left[ f(y) - f(x) \right] d\sigma(y) \right|$$

*is bounded as a map from $H^{1,p}(\Gamma)$ to $L^p(\Gamma)$ for $1 < p < \infty$, where $\Gamma$ is a graph of a Lipschitz function $\phi : \mathbb{R}^{n-1} \to \mathbb{R}$. Here $r(x,y)$ means the geodesic distance on $\Gamma$. In particular, the linear operator*

(A.30) $$K^\phi f(x) = P.V. \int_\Gamma k(x-y) g(y) \left[ f(y) - f(x) \right] d\sigma(y)$$

is well defined and bounded as a map from $H^{1,p}(\Gamma)$ to $L^p(\Gamma)$. Actually, we have an estimate

$$\|K^\phi f\|_{L^p(\Gamma)} \leq C(n,\Gamma)\|k\|_{C^N(S^{n-1})}\|g\|_{L^\infty}\|f\|_{H^{1,p}(\Gamma)}. \tag{A.32}$$

Moreover, $\phi \mapsto K^\phi$, provided we look at $K^\phi$ as an operator on $\mathbb{R}^{n-1}$, is a continuous function from $\mathrm{Lip}(\mathbb{R}^{n-1})$ to $\mathcal{L}(H^{1,p}(\mathbb{R}^{n-1}), L^p(\mathbb{R}^{n-1}))$.

PROOF. Case $n = 2$ can be established using Theorem 4 of [**3**] following the approach taken in the paper of Coifman, David and Meyer [**4**]. Namely with $x, y \in \Gamma \subset \mathbb{R}^2$ we get that

$$k(x-y) = k(x_1 - y_1, \phi(x_1) - \phi(y_1)) = \frac{1}{(x_1-y_1)^2} k\left(1, \frac{\phi(x_1) - \phi(y_1)}{x_1 - y_1}\right), \tag{A.33}$$

where $x_1, y_1 \in \mathbb{R}$. That is, our operator $K^*$ will become

$$K^* f(x) = \sup_{\varepsilon > 0} \left| \int_{|x_1 - y_1| > \varepsilon} \frac{f(y_1) - f(x_1)}{|y_1 - x_1|} k\left(1, \frac{\phi(x_1) - \phi(y_1)}{x_1 - y_1}\right) \right.$$
$$\left. \frac{1}{|y_1 - x_1|} g(y_1) \sqrt{1 + (\phi'(y_1))^2} \, dy_1 \right|. \tag{A.34}$$

The functions $f$ and $g$ are now seen as functions on $H^{1,p}(\mathbb{R})$, $L^\infty(\mathbb{R})$, respectively. This is exactly what is needed for Theorem 4 of [**3**] with one exception. Namely, in [**3**] it is required that $G(z) = k(1, z)$ is holomorphic in $z$, whereas our function is only $C^N$. However, later results, in particular [**4**], [**8**] and [**9**], showed that holomorphicity is not necessary and can be replaced by sufficient smoothness. The main idea of the proof is to write $G$ in terms of its Fourier transformation. (I would like to thank professor *Alan McIntosh* for this hint). Thus we have

$$\|K^*(f,g)\|_{L^p(\Gamma)} \leq C(n,\Gamma)\|k\|_{C^N(S^{n-1})}\|g\|_{L^\infty}\|f\|_{H^{1,p}(\Gamma)}. \tag{A.35}$$

Using the standard method of rotation we get same estimate as (A.35) also for $n > 2$. Then, (A.35) gives (A.32). Finally, the continuous dependence of $\phi \mapsto K^\phi$ for $\phi \in C(\mathbb{R}^{n-1})$ follows from an argument similar to one used in Proposition A.3, namely using the procedure outlined above one can show that also

$$T(f, g, A)(x) = \mathrm{P.V.} \int_\Gamma k(x-y) g(y) \left[f(y) - f(x)\right] \left[A(y) - A(x)\right] \, d\sigma(y) \tag{A.36}$$

is well defined and

$$\|T(f, g, A)\|_{L^p(\Gamma)} \leq C(n,\Gamma)\|k\|_{C^N(S^{n-1})}\|\nabla A\|_{L^\infty}\|g\|_{L^\infty}\|f\|_{H^{1,p}(\Gamma)}, \tag{A.37}$$

for $A$ Lipschitz, $f$, $g$ as before, and $k$ odd, homogeneous of degree $-(n+1)$, and smooth enough. This corresponds to Proposition A.2 which was needed in the proof of Proposition A.3. □

Now, by same technique as in Proposition A.4 (spherical decomposition), we can also get:

PROPOSITION A.10. *Let $\Gamma$ be a graph of a Lipschitz function $\phi : \mathbb{R}^{n-1} \to \mathbb{R}$. There is $M = M(n)$ such that the following holds. Let $b(x, z)$ be even in $z$, homogeneous of degree $-n$ in $z$, and assume also that $D_z^\alpha b(x, z)$ is continuous and bounded on $\mathbb{R}^n \times S^{n-1}$ for $|\alpha| \leq M$. Then the operator*

$$(A.38) \qquad B^\phi f(x) = \int_\Gamma b(x, x - y) g(y) \left[ f(y) - f(x) \right] \, d\sigma(y)$$

*is bounded as a map from $H^{1,p}(\Gamma)$ to $L^p(\Gamma)$ for $1 < p < \infty$. In particular, we have an estimate*

$$(A.39) \qquad \|B^\phi f\|_{L^p(\Gamma)} \leq C(n, \Gamma) \sup_{|\alpha| \leq M} \|D_z^\alpha b(x, z)\|_{L^\infty(\mathbb{R}^n \times S^{n-1})} \|g\|_{L^\infty} \|f\|_{H^{1,p}(\Gamma)}.$$

*Moreover, if we look at $B^\phi$ as an operator on $\mathbb{R}^{n-1}$, then $\phi \mapsto B^\phi$ is a continuous function from $Lip(\mathbb{R}^{n-1})$ to $\mathcal{L}(H^{1,p}(\mathbb{R}^{n-1}), L^p(\mathbb{R}^{n-1}))$.*

PROPOSITION A.11. *There is $N = N(n)$ such that the following holds. Let $k(.)$ be even, homogeneous of degree $-n$, and $k \in C^N(\mathbb{R}^n \setminus \{0\})$. Then the operator*

$$(A.40) \qquad \mathcal{K}f(x, x_0) = \int_\Gamma k(x - y) g(y) \left[ f(y) - f(x_0) \right] \, d\sigma(y),$$

*defined for $x \notin \Gamma$ and $x_0 \in \Gamma$, satisfies the following estimate on nontangential maximal function:*

$$(A.41) \qquad \| \sup_{x \in \gamma(x_0)} |\mathcal{K}f(x, x_0)| \|_{L^p(\Gamma)} \leq C \|g\|_{L^\infty} \|f\|_{H^{1,p}(\Gamma)}$$

*for any $1 < p < \infty$. Here, $\gamma(x_0)$ is the nontangential approach region to a point $x_0$.*

PROOF. We can write (A.40) as follows. Let $x \in \gamma(x_0)$ and $\eta > 0$.

$$|\mathcal{K}f(x, x_0)| \leq \int_{\Gamma \cap \{|y - x_0| \leq \eta\}} |k(x - y)| |g(y)| |f(y) - f(x_0)| \, d\sigma(y) +$$
$$+ \int_{\Gamma \cap \{|y - x_0| > \eta\}} |k(x - y) - k(x_0 - y)| |g(y)| |f(y) - f(x_0)| \, d\sigma(y) +$$
$$(A.42) \qquad + \left| \int_{\Gamma \cap \{|y - x_0| > \eta\}} k(x_0 - y) g(y) \left[ f(y) - f(x_0) \right] \, d\sigma(y) \right|.$$

Term $|f(y) - f(x_0)|$ in first two integrals can be dominated by $\overline{f}^*(x_0) |y - x_0|$ where

$$(A.43) \qquad \overline{f}^*(x_0) = \sup_{r > 0} \frac{1}{r^{n-1}} \int_{\Gamma \cap \{|y - x_0| \leq r\}} |\nabla f|(y) \, d\sigma(y).$$

Now, in the first integral, since $x$ approaches $x_0$ from inside $\gamma(x_0)$, we have that $|x - y| \approx |x - x_0| + |y - x_0|$. It follows that

$$(A.44) \qquad |k(x - y)| |y - x_0| \leq C \frac{|y - x_0|}{|x - x_0|^n + |y - x_0|^n}.$$

In particular, if we take $\eta = |x - x_0|$, then we can bound the first term of (A.42) by $C\overline{f}^*(x_0)$. For the second term we have

$$(A.45) \qquad |k(x - y) - k(x_0 - y)| \leq C \frac{|x - x_0|}{|x_0 - y|^{n+1}},$$

using homogeneity, and the fact that for $|y - x_0| > \eta = |x - x_0|$ we have $|x - y| \approx |x_0 - y|$. This gives that the second integral can be also estimated by $C\overline{f}^*(x_0)$. Finally, the third integral we estimate by

$$(A.46) \qquad K^* f(x_0) = \sup_{\eta > 0} \left| \int_{\Gamma \cap \{|y - x_0| > \eta\}} k(x_0 - y) g(y) \left[ f(y) - f(x_0) \right] d\sigma(y) \right|,$$

which belongs to $L^p(\Gamma)$ by Proposition A.9. Since also $\|\overline{f}^*\|_{L^p(\Gamma)} \leq C \|\nabla f\|_{L^p(\Gamma)}$, all terms in (A.42) are bounded by $C\|f\|_{H^{1,p}(\Gamma)}$. From this the claim follows. $\square$

Now using the spherical decomposition we can establish:

PROPOSITION A.12. *There is $M = M(n)$ such that the following holds. Let $b(x, z)$ be even in $z$, homogeneous of degree $-n$ in $z$, and assume that $D_z^\alpha b(x, z)$ is continuous and bounded on $\mathbb{R}^n \times S^{n-1}$ for $|\alpha| \leq M$. Then the operator*

$$(A.47) \qquad \mathcal{B}f(x, x_0) = \int_\Gamma b(x, x - y) g(y) \left[ f(y) - f(x_0) \right] d\sigma(y),$$

*defined for $x \notin \Gamma$ and $x_0 \in \Gamma$, satisfies the following estimate on nontangential maximal function:*

$$(A.48) \qquad \|\sup_{x \in \gamma(x_0)} |\mathcal{B}f(x, x_0)| \|_{L^p(\Gamma)} \leq C \|g\|_{L^\infty} \|f\|_{H^{1,p}(\Gamma)}$$

*for $1 < p < \infty$. Here, $\gamma(x_0)$ is as before the nontangential approach region to a point $x_0$.*

Now we look again at Hardy spaces to get an analogue of Proposition A.12. First, we establish a lemma similar to Proposition A.11, with one additional commutator term.

LEMMA A.13. *Let $1 < p < \infty$. There is $N = N(n)$ such that the following holds. Let $k(.)$ be odd, homogeneous of degree $-(n+1)$, and $k \in C^N(\mathbb{R}^n \setminus \{0\})$. Consider the operator*

$$(A.49) \qquad \mathcal{K}f(x, x_0) = \int_\Gamma k(x - y)(A(y') - A(x')) g(y) \left[ f(y) - f(x_0) \right] d\sigma(y),$$

*defined for $x \notin \Gamma$, $x_0 \in \Gamma$, $A : \mathbb{R}^{n-1} \to \mathbb{R}$ Lipschitz and $f$, $g$ as above. Here, we use the notation $x = (x', x_n) \in \mathbb{R}^n$, with $x' \in \mathbb{R}^{n-1}$ and $x_n \in \mathbb{R}$. We have the following estimate on the nontangential maximal function:*

$$(A.50) \qquad \|\sup_{x \in \gamma(x_0)} |\mathcal{K}f(x, x_0)| \|_{L^p(\Gamma)} \leq C \|g\|_{L^\infty} \|\nabla A\|_{L^\infty} \|f\|_{H^{1,p}(\Gamma)}.$$

PROOF. Since the proof of this lemma is virtually same as the proof of Propositions A.9 and A.11, we skip it. $\square$

In particular, taking $k(z)$ even homogenous of degree $-n$ we get that Lemma A.11 applies to $\frac{\partial}{\partial z_n} k(z)$. Lemma A.11 yields an uniform estimate (independent of $\varepsilon$) of the type

$$(A.51) \qquad \|\sup_{x \in \gamma(x_0)} |T^\varepsilon f(x, x_0)| \|_{L^p(\Gamma)} \leq C \|g\|_{L^\infty} \|\nabla A\|_{L^\infty} \|f\|_{H^{1,p}(\Gamma)},$$

where

$$T^\varepsilon f(x, x_0) = \int_\Gamma \frac{\partial}{\partial z_n} k(x - y + e_n \varepsilon (A(x') - A(y'))) \times$$
(A.52)
$$g(y)[f(y) - f(x_0)][A(y') - A(x')] \, d\sigma(y).$$

Here $e_n = (0, \ldots, 0, 1) \in \mathbb{R}^n$ is a unit vector. If $\varepsilon > 0$, (A.52) simply means that we do not integrate over the surface $\Gamma$, but over a surface which we call $\Gamma^\varepsilon$ obtained from $\Gamma$ by a shift by $\varepsilon A$ in the $e_n$ direction. Because of this, the domain of this operator is $\mathbb{R}^n \setminus \Gamma^\varepsilon$, i.e., for $x_0$ fixed we take nontangential approach region $\gamma(x_0 + e_n \varepsilon A(x_0))$.

Imagine now that we have two Lipschitz surfaces $\Gamma_1$, $\Gamma_0$ in $\mathbb{R}^n$, given by two Lipschitz functions $\phi^0$, $\phi^1$. For a brief period we equip $\Gamma_1$ by the measure taken from $\Gamma_0$ by vertical projection. Hence we can identify functions on $L^p(\Gamma_1)$ with $L^p(\Gamma_0)$. We put $A = \phi_1 - \phi_0$. It follows that the surface $\Gamma^\varepsilon$, defined above, is simply a graph of the Lipschitz function $(1 - \varepsilon)\phi^0 + \varepsilon \phi^1$.

Take the kernel $k$ as in Proposition A.11. We want to estimate the difference

(A.53)
$$\| \sup_{x \in \gamma(x_0)} |(\mathcal{K}^1 - \mathcal{K}^0) f(x, x_0)| \|_{L^p(\Gamma)}.$$

Here $\mathcal{K}^\varepsilon$ for $\varepsilon \in [0, 1]$ is an operator defined by (A.49) corresponding to $\Gamma_\varepsilon$, with measure taken from $\Gamma_0$. (A.53) is understood as in Proposition A.6. That is, for a fixed point $x_0 \in \Gamma_0$ and its nontangential approach region $\gamma(x_0)$, there is exactly one point $\widetilde{x_0} \in \Gamma_1$ such that the first $n - 1$ coordinates of $x_0$ and $\widetilde{x_0}$ are equal. We compare the maximal operator at $x_0$ with maximal operator at $\widetilde{x_0}$ by shifting vertically the whole domain of $\mathcal{K}^1 f(x, \widetilde{x_0})$ such that the points $x_0$ and $\widetilde{x_0}$ coincide. By doing this, we achieve that the nontangential approach regions $\gamma(x_0)$ and $\gamma(\widetilde{x_0})$ coincide, hence we can compute $((\mathcal{K}^1 - \mathcal{K}^0) f(x, x_0))^*$ at $x_0$. Naturally, the considered shift is different for different $x_0$. The precise way to write (A.52) is

(A.54)
$$\| \sup_{x \in \gamma(x_0)} |(\mathcal{K}^1 f(x + e_n A(x_0'), x_0 + e_n A(x_0')) - \mathcal{K}^0 f(x, x_0)| \|_{L^p(\Gamma)}.$$

However, for simplicity we stick to the notation introduced in (A.53). In a special case, when $\phi_0 = \phi_1 + c$ for some constant $c$, we get that the above shift is by $c$, and $((\mathcal{K}^1 - \mathcal{K}^0) f(x, x_0))^* = 0$. Now (A.51) guarantees that

(A.55)
$$\| \sup_{x \in \gamma(x_0)} |\tfrac{d}{d\varepsilon} \mathcal{K}^\varepsilon f(x, x_0)| \|_{L^p(\Gamma)} \leq C \|g\|_{L^\infty} \|\nabla A\|_{L^\infty} \|f\|_{H^{1,p}(\Gamma)}.$$

From this estimate for (A.53) follows. The final point is, that changing the measure on $\Gamma_1$ back to its natural surface measure produces an additional term estimable easily, since it could be included into function $g$. This establishes:

LEMMA A.14. *Let $1 < p < \infty$. There is $N = N(n)$ such that the following holds. Let $k(.)$ be even, homogeneous of degree $-n$, and $k \in C^N(\mathbb{R}^n \setminus \{0\})$. Consider two surfaces in $\mathbb{R}^n$, $\Gamma_0$, $\Gamma_1$ given by Lipschitz functions $\phi^0$, $\phi^1$ and on each of them an operator*

(A.56)
$$\mathcal{K}^i f(x, x_0) = \int_{\Gamma^i} k(x - y) g(y) [f(y) - f(x_0)] \, d\sigma^i(y),$$

where $i \in \{0,1\}$, $x \notin \Gamma_i$, $x_0 \in \Gamma_i$. If $\gamma(x_0)$ is a family of nontangential approach regions for $x_0 \in \Gamma_0$, then the following estimate on the nontangential maximal function for the diference $(\mathcal{K}^1 - \mathcal{K}^0)f(x, x_0)$ holds.

$$(A.57) \quad \|\sup_{x \in \gamma(x_0)} |(\mathcal{K}^1 - \mathcal{K}^0)f(x, x_0)|\|_{L^p(\Gamma_0)} \leq C \|\nabla(\phi^1 - \phi^0)\|_{L^\infty} \|g\|_{L^\infty} \|f\|_{H^{1,p}(\Gamma_0)}.$$

Finally, the previous lemma is used to establish the desired result for Hardy spaces.

PROPOSITION A.15. *Let $(n-1)/n < p \leq 1$. There exists $N = N(n)$ such that the following holds. Let $k \in C^N(\mathbb{R}^n \setminus 0)$ be even, homogeneous of degree $-n$, and consider two surfaces in $\mathbb{R}^n$, $\Gamma_0$, $\Gamma_1$ given by Lipschitz functions $\phi^0$, $\phi^1$. For $i = 0, 1$ consider the operator operator*

$$(A.58) \quad \mathcal{K}^i f(x, x_0) = \int_{\Gamma^i} k(x-y) g(y) \left[ f(y) - f(x_0) \right] d\sigma^i(y),$$

*defined for $x \notin \Gamma_i$ and $x_0 \in \Gamma_i$. Then for each $i$, $\mathcal{K}^i$ is well defined for $f \in H^{1,p}(\Gamma_i)$. Moreover, if $\gamma(x_0)$ is a family of nontangential approach regions for $x_0 \in \Gamma_0$, then the following estimate on the nontangential maximal function for the difference $(\mathcal{K}^1 f - \mathcal{K}^0 f)(x, x_0)$ holds:*

$$(A.59)$$
$$\|\sup_{x \in \gamma(x_0)} |(\mathcal{K}^1 f - \mathcal{K}^0 f)(x, x_0)|\|_{L^p(\Gamma_0)} \leq C \omega(\|\phi^1 - \phi^0\|_{Lip(\mathbb{R}^{n-1})}) \|g\|_{L^\infty} \|f\|_{H^{1,p}(\Gamma_0)},$$

*where $\omega$ is as before a decreasing function continuous at 0 with $\omega(0) = 0$ (modulus of continuity).*

PROOF. As in [23], we just have to consider normalized vector $p$-atoms, i.e., take $f \in H^{1,p}(\Gamma_0)$ satisfying

$$(A.60) \quad \operatorname{supp} \nabla f \subset B_1(0) \cap \Gamma_0, \quad \|\nabla f\|_{L^\infty(\Gamma_0)} \leq 1, \quad \int_{\Gamma_0} \nabla f \, d\sigma_0 = 0,$$

and $0 \in \Gamma_0$. Here the third condition is actually not necessary and follows from the fact that outside $B_1(0)$ the function $f$ must be constant. Hence, we can look at $f$ as a function on another Lipschitz curve $\Gamma_1$, where it also satisfies conditions similar to (A.60). (With possibly larger estimate on the $L^\infty$ norm of $\nabla f$).

Now, for any $\varepsilon > 0$ there is $R$ big such that

$$(A.61) \quad \begin{aligned} |x_0| \geq R &\Longrightarrow |\mathcal{K}^1 f(x, x_0)| \leq \varepsilon \|g\|_{L^\infty} |x|^{-n}, \\ &\quad |\mathcal{K}^0 f(x, x_0)| \leq \varepsilon \|g\|_{L^\infty} |x|^{-n}. \end{aligned}$$

On $B_R(0)$ the $L^2$ theory from the previous lemma leads to an estimate

$$(A.62) \quad \begin{aligned} \int_{\Gamma_0 \cap B_R(0)} [\sup_{x \in \gamma(x_0)} (\mathcal{K}^1 f - \mathcal{K}^0 f)(x, x_0)]^p \|g\|_{L^\infty} \, d\sigma_0(x_0) \leq \\ \leq C_p \, \omega(\|\phi_1 - \phi_0\|_{Lip(\mathbb{R}^{n-1})}) \|g\|_{L^\infty}. \end{aligned}$$

Combining (A.61), (A.62) yields (A.59) for vector atoms. Density argument brings then (A.59) for any $f \in H^{1,p}(\Gamma_0)$. □

Finally, by spherical decomposition we obtain the following:

PROPOSITION A.16. *Let $(n-1)/n < p \le 1$. There exists $M = M(n)$ such that the following holds. Let $b(x, z)$ be even in $z$, homogeneous of degree $-n$ in $z$, and assume that $D_z^\alpha b(x, z)$ is continuous and bounded on $\mathbb{R}^n \times S^{n-1}$ for $|\alpha| \le M$. Consider two surfaces in $\mathbb{R}^n$, $\Gamma_0$, $\Gamma_1$ given by Lipschitz functions $\phi^0$, $\phi^1$, and on each of them an operator*

$$\mathcal{B}^i f(x, x_0) = \int_{\Gamma^i} b(x, x - y) g(y) \left[ f(y) - f(x_0) \right] d\sigma^i(y), \tag{A.63}$$

*where $i \in \{0, 1\}$, $x \notin \Gamma_i$, $x_0 \in \Gamma_i$. We claim that each of them is well defined on $H^{1,p}(\Gamma_i)$, and*

$$\| \sup_{x \in \gamma(x_0)} |\mathcal{B}^i f(x, x_0)| \|_{L^p(\Gamma_i)} \le C \|g\|_{L^\infty} \|f\|_{H^{1,p}(\Gamma_i)}, \tag{A.64}$$

*where $\gamma(x_0)$ is a family of nontangential approach regions for $x_0 \in \Gamma_0$.*

*Also for the difference $(\mathcal{B}^1 f - \mathcal{B}^0 f)(x, x_0)$ understood in a sense explained above we get:*
(A.65)
$$\| \sup_{x \in \gamma(x_0)} |(\mathcal{B}^1 f - \mathcal{B}^0 f)(x, x_0)| \|_{L^p(\Gamma_0)} \le C \omega(\|\phi^1 - \phi^0\|_{Lip(\mathbb{R}^{n-1})}) \|g\|_{L^\infty} \|f\|_{H^{1,p}(\Gamma_0)},$$

*where $\omega$ is as before a decreasing function continuous at 0 with $\omega(0) = 0$.*

At the end, we would like to present a result somewhat similar to Proposition A.10 but for less singular kernels.

PROPOSITION A.17. *Let $\Gamma$ be as before a bounded graph of a Lipschitz surface given by $\phi : \mathbb{R}^{n-1} \to \mathbb{R}$. Let the kernel $b(x, y)$ be a continuous function off the diagonal $\{x = y\}$, and let for any $\delta > 0$ we have the following growth condition on $b(x, z)$ near the diagonal:*

$$|b(x, y)| \le C_\delta |x - y|^{-(n-1+\delta)}. \tag{A.66}$$

*Then for any $1 \le p \le \infty$, the operator*

$$Bf(x) = \int_\Gamma b(x, y) \left[ f(y) - f(x) \right] d\sigma(y) \tag{A.67}$$

*is bounded and maps $H^{\varepsilon, p}(\Gamma)$ to $L^p(\Gamma)$, for any $\varepsilon > 0$. In particular, we have an estimate*

$$\|Bf\|_{L^p(\Gamma)} \le C \|f\|_{H^{\varepsilon, p}(\Gamma)}. \tag{A.68}$$

PROOF. Clearly, if we establish the result for $p = 1$ and $p = \infty$, by interpolation we have it for all $p$. Consider first the case $p = \infty$. If $f \in H^{\varepsilon, \infty}(\Gamma)$, then clearly for some $\delta > 0$, $f \in C^{2\delta}(\Gamma)$. This allows us to estimate:

$$\begin{aligned}|Bf(x)| &\le \int_\Gamma |b(x, y)| |x - y|^{2\delta} \frac{|f(y) - f(x)|}{|y - x|^{2\delta}} d\sigma(y) \\ &\le C \|f\|_{C^{2\delta}(\Gamma)} \int_\Gamma |b(x, y)| |x - y|^{2\delta} d\sigma(y).\end{aligned} \tag{A.69}$$

The last integral in (A.69) is finite due to (A.66). We get

$$|Bf(x)| \le C(\Gamma, b) \|f\|_{C^{2\delta}(\Gamma)} \le C(\Gamma, b, \delta) \|f\|_{H^{\varepsilon, \infty}(\Gamma)}. \tag{A.70}$$

Now, consider $p = 1$. If $0 < \delta < 1$, then the Besov space $B^1_\delta(\Gamma)$ can be characterized as the space of all functions $f$ for which the following number (norm)

$$\text{(A.71)} \qquad \|f\|_{B^1_\delta(\Gamma)} = \|f\|_{L^1(\Gamma)} + \int_\Gamma \int_\Gamma \frac{|f(y) - f(x)|}{|x - y|^{n-1+\delta}} \, d\sigma(x) \, d\sigma(y)$$

is finite. In particular, if $f \in H^{\varepsilon,1}$, then $f \in B^1_\delta$ for any $0 < \delta < \varepsilon$. Hence if we take $\delta = \frac{1}{2}\varepsilon$ we get:

$$\|Bf\|_{L^1(\Gamma)} = \int_\Gamma \left| \int_\Gamma b(x,y) \left[ f(y) - f(x) \right] \, d\sigma(y) \right| \, d\sigma(x)$$

$$\text{(A.72)} \qquad \leq \int_\Gamma \int_\Gamma |b(x,y)| |f(y) - f(x)| \, d\sigma(y) \, d\sigma(x)$$

$$\leq C_\delta \int_\Gamma \int_\Gamma \frac{|f(y) - f(x)|}{|x - y|^{n-1+\delta}} \, d\sigma(y) \, d\sigma(x) \leq C(\Gamma, b, \delta) \|f\|_{H^{\varepsilon,1}(\Gamma)}.$$

We used (A.71) to estimate the last term. This concludes our proof. $\square$

APPENDIX B

# One Result on the Maximal Operator

In this appendix we study the actions of $(\Delta - V)^{-1}$ on functions $f$ whose maximal operator $f^*$ belongs to $L^p(\partial\Omega)$. The main result, Theorem B.10 is crucial in Chapter 8 of this work. The approach we present here has been generalized recently, see [12] and [13] for reference.

The whole proof is highly technical, for this reason, we split it into several auxiliary lemmas and Propositions. The general setting of this appendix is exactly same as before, i.e., $M$ is an $n$-dimensional Riemannian manifold with metric tensor of regularity $C^\alpha$, $\Omega \subset M$ is a connected domain with Lipschitz boundary, $L = \Delta - V$ is the considered operator on $M$ with an inverse $L^{-1}$ which can be written as

$$\text{(B.1)} \qquad L^{-1}u(x) = \int_M E(x,y)u(y)\,d\text{Vol}(y).$$

Recall, the estimates on the kernel of $E$ given by (1.21), (1.22) and (1.23). In particular, it follows that

$$\text{(B.2)} \qquad |E(x,y)| \leq \frac{C}{|x-y|^{n-2}},$$

where $|x-y|$ means the geodesic distance of the points $x$ and $y$ on $M$ (slightly abusing the notation).

LEMMA B.1. *Let $x \in M$ be an arbitrary point and $r > 0$. Consider a geodesic ball $B_r(x)$ of radius $r$ around $x$ and assume that $f \in L^\infty(M)$ is a given function with support in $B_r(x)$, bounded in absolute value by one. Let $u$ be the solution to*

$$\text{(B.3)} \qquad Lu = f \quad \text{on } M, \quad \text{i.e.,} \quad u = L^{-1}f.$$

*There exists a constant $C > 0$ independent on $f$ and $x$ such that for any $y \in M$*

$$\text{(B.4)} \qquad |u(y)| \leq C\frac{r^{n-1}}{(r+|x-y|)^{n-3}}.$$

PROOF. We prove (B.4) in two steps. First we consider $y \in B_{2r}(x)$. We estimate $|u(y)|$ using (B.1). Assume for simplicity that $r > 0$ is small enough, such that we can consider just one coordinate chart centered at $x$ that contains the ball $B_{2r}(x)$. In this chart we can also assume that $x$ is the origin. We integrate over $(n-1)$-dimensional shells $S_\rho = \partial B_\rho(y)$ centered at $y$. Simple estimate using (B.2) gives

$$\text{(B.5)} \qquad |u(y)| \leq C\int_0^{3r}\int_{S_\rho} \frac{1}{|z-y|^{n-2}}\,d\sigma(z)\,d\rho.$$

Since the surface area of $S_\rho$ is of the order of $\rho^{n-1}$, (B.5) gives us

$$|u(y)| \leq C \int_0^{3r} \rho \, d\rho \leq C r^2. \tag{B.6}$$

By possibly enlarging the constant $C$ in (B.6) we can see that (B.6) and (B.4) are equivalent for $y \in B_{2r}(x)$.

Now we consider $y$ outside the ball $B_{2r}(x)$. Denote by $\varepsilon$ the distance between $y$ and $B_r(x)$. We integrate the same way we did in the first part over $S_\rho$. However, it is clear that in this case $S_\rho$ intersects the support of $f$ only for $\rho \geq \varepsilon$. Moreover, the surface measure of such intersection can be estimated by $C r^{n-1}$. This leads to

$$\begin{aligned} |u(y)| &\leq C \int_\varepsilon^\infty \int_{S_\rho \cap B_r(x)} \frac{1}{|z-y|^{n-2}} \, d\sigma(z) \, d\rho \leq \\ &\leq C \int_\varepsilon^\infty r^{n-1} \frac{1}{\rho^{n-2}} \, d\rho \leq C \frac{r^{n-1}}{\varepsilon^{n-3}}. \end{aligned} \tag{B.7}$$

Now since $\varepsilon \approx |x-y| - r$ we get that for $|x-y| \geq 2r$ (i.e., $y \notin B_{2r}(x)$)

$$\varepsilon \approx |x-y| + r. \tag{B.8}$$

This implies that the estimate (B.2) works for such $y$. □

Our next step is an estimate on the $L^p(\partial\Omega)$ norm of $(L^{-1}f)^*$, for $f$ as in the previous lemma.

PROPOSITION B.2. *Assume that $r > 0$ is small. Let $f \in L^\infty(M)$ be a function on $M$ bounded in absolute value by one with support in $B_r(x) \cap \Omega$, where $x$ is a point on the boundary $\partial\Omega$. Let $u = L^{-1}f$ be as before. Then for $1 \leq p < (n-1)/(n-3)$, the following estimate on the $L^p$ norm of the nontangential maximal operator $u^*$ holds:*

$$\|u^*\|_{L^p(\partial\Omega)} \leq C r^{n-1}. \tag{B.9}$$

PROOF. Since $r > 0$ is small, we can find a small neighborhood $U$ of $x$ such that in this neighborhood there are smooth local coordinates in which

$$U \cap \Omega = \{x = (x', x_n) \in U : x_n > \varphi(x')\}, \tag{B.10}$$

where $\varphi$ is a Lipschitz function with Lipschitz constant bounded by $L$. Here $L$ does not depend on the chosen point $x \in \partial\Omega$. Take a collection of nontangential approach regions $\gamma(z)$ to any point boundary point $z = (z', \varphi(z')) \in \partial\Omega$ whose vertex at $z$ is sharp enough. Namely, we require that any half-ray with vertex at $z$ that lies whole in $\gamma(z)$ has "steepness" (absolute value of its slope) at least $2L$.

Given such collection, it follows that we can find a universal constant $k$ (independent of $r$), such that we can split points $z \in \partial\Omega$ into two distinct sets. If $z = (z', \varphi(z')) \in \partial\Omega$ and $|z' - x'| \leq kr$ then $\gamma(z)$ might intersect $B_r(x)$. At such point we estimate $u^*(y)$ by $C r^2$, by (B.4). On the other hand if $|z' - x'| > kr$ then the distance between any point $w \in \gamma(z)$ and $x$ is greater or equal to $\frac{1}{k}|z' - x'|$. This means that for such $z$ we can estimate $u^*(y)$ by

$$u^*(y) \leq C \frac{r^{n-1}}{(r + k^{-1}|z' - x'|)^{n-3}}. \tag{B.11}$$

Now we take the $L^p$ norm of $u^*$. On $B_{kr}(x) \cap \partial\Omega$ we get

$$\text{(B.12)} \qquad \int_{B_{kr}(x)\cap\partial\Omega} (u^*(y))^p \, d\sigma(y) \leq C r^{n-1} r^{2p} = C r^{n+2p-1}.$$

Similarly, off $B_{kr}(x) \cap \partial\Omega$ we get

$$\int_{\partial\Omega \setminus B_{kr}(x)} (u^*(y))^p \, d\sigma(y) \leq C \int_{kr}^{A} \int_{y \in S_\rho} \left( \frac{r^{n-1}}{(r+k^{-1}\rho)^{n-3}} \right)^p d\sigma(y) \, d\rho \leq$$

$$\text{(B.13)} \qquad \leq C r^{p(n-1)} \int_0^A \frac{\rho^{n-2}}{(r+k^{-1}\rho)^{p(n-3)}} \, d\rho.$$

Here $S_\rho$ is a $(n-2)$-dimensional shell defined by

$$\text{(B.14)} \qquad S_\rho = \partial\Omega \cap \partial B_\rho(x).$$

In (B.13) we have also used that $(n-2)$-dimensional surface area of such shell is of the order $\rho^{n-2}$. We further estimate the integral in (B.13):

$$\text{(B.14)} \qquad \int_0^A \frac{\rho^{n-2}}{(r+k^{-1}\rho)^{p(n-3)}} \, d\rho \leq C \int_0^A \rho^{n-2-p(n-3)} \, d\rho.$$

If $p < (n-1)/(n-3)$ then $n-2-p(n-3) > -1$, hence (B.14) is finite (and independent of $r$). Finally, we put (B.12), (B.13) and (B.14) together to get

$$\text{(B.15)} \quad \|u^*\|_{L^p(\partial\Omega)} \leq C(r^{n-1+2p} + r^{p(n-1)})^{1/p} \leq C(r^{2+(n-1)/p} + r^{n-1}) \leq C r^{n-1}.$$

In the final estimate in (B.15) we used again that $p < (n-1)/(n-3)$, and therefore $2 + (n-1)/p > n-1$. This concludes our proof. $\square$

Let $z = (z', z_n)$ be any point in the coordinate chart (B.10). We put

$$\text{(B.16)} \qquad \widetilde{\gamma(z)} = \{w = (w', w_n) \in U : w_n < z_n \text{ and } |z' - w'| < 2L|z_n - w_n|\}.$$

Here $L$ is as before a bound on the Lipschitz constant of $\partial\Omega$. So our region $\widetilde{\gamma(z)}$ is an open downward opening cone with vertex at $z$. If we compare $\gamma(z)$ and $\widetilde{\gamma(z)}$ we can see that these two cones are symmetric with respect to the hyperplane $x_n = z_n$ in $U$.

DEFINITION B.3. Consider a coordinate chart (B.10) and a set $A \subset \partial\Omega$ on this chart which is open in $\partial\Omega$. We say that a set $\mathcal{A} \subset \Omega$ is a $P$-image of $A$ and write

$$\text{(B.17)} \qquad \mathcal{A} = \text{Pim}(A),$$

provided the set $\mathcal{A}$ satisfies the following conditions:

(a) The set $\{z = (z', \varphi(z')) \in \partial\Omega : \exists w = (z', w_n) \in \mathcal{A}\}$ (a projection of $\mathcal{A}$ onto $\partial\Omega$) is $A$.

(b) $z = (z', z_n) \in \mathcal{A}$ if and only if $z \in \Omega$ and $\widetilde{\gamma(z)} \cap \partial\Omega \subset A$.

REMARK. The property (a) follows from (b). It also follows, that if $z \in \mathcal{A}$ then $\widetilde{\gamma(z)} \cap \Omega \subset \mathcal{A}$.

Now we establish a connection between the set $\mathcal{A}$ from the previous definition and Proposition B.2.

PROPOSITION B.4. *Let the set $A$ be as in the definition B.3 and $\mathcal{A} = Pim(A)$. Consider $u = L^{-1}f$, where $f \in L^\infty(M)$ is a function on $M$ bounded in absolute value by one, with support in $\mathcal{A}$. For any $1 \leq p < (n-1)/(n-3)$, there is a constant $C_p$ (independent on $A$) such that*

$$\|u^*\|_{L^p(\partial\Omega)} \leq C\sigma(A), \tag{B.18}$$

*where $\sigma(A)$ is the $(n-1)$-dimensional (surface) area of $A$ on $\partial\Omega$.*

PROOF. We will do our proof in several steps. Given any point $x \in A$, we assign to it a positive "height" number $h(x) > 0$ as follows. In the coordinate chart we can write $x$ as $(x', x_n)$. Define

$$h(x) = \sup\{t \in \mathbb{R}^+ : y = (x', x_n + t) \in \mathcal{A}\}. \tag{B.19}$$

Now clearly, $h(x) > 0$ for $x \in A$, and the whole set $\Omega \cap \widetilde{\gamma}(x', x_n + h(x))$ is contained in $\mathcal{A}$. Here, $\widetilde{\gamma}(.)$ is the cone defined by (B.16). The set

$$V = \partial\Omega \cap \widetilde{\gamma}(x', x_n + h(x)) \tag{B.20}$$

is an open neighborhood of $x$ on $\partial\Omega$. Also, $V \subset A$. Using the fact that the surface $\partial\Omega$ is Lipschitz we can also establish a relation between $h(x)$ and $\sigma(V)$. Namely, there are two positive constants $c_1$ and $c_2$ depending only on the Lipschitz character of $\partial\Omega$ such that

$$c_1(h(x))^{n-1} \leq \sigma(V) \leq c_2(h(x))^{n-1}. \tag{B.21}$$

An immediate consequence of this observation is, that the number

$$H = \sup_{x \in A} h(x) \tag{B.22}$$

is finite, since the surface measure of $A$ is finite. Denote by $\mathcal{H}$ the hyperplane

$$\{x = (x', 0) : x' \in \mathbb{R}^{n-1}\}. \tag{B.22}$$

On the chart (B.10) we consider a projection $P : U \to \mathcal{H}$ which assigns to any point $x = (x', x_n)$ the point $P(x) = (x', 0) \in \mathcal{H}$.

We divide $\mathcal{H}$ into a union of $(n-1)$-dimensional disjoint cubes with sides of length $2H/L$. Let us denote this collection of cubes by $\mathcal{C}_1$. Let $\mathcal{D}_1 \subset \mathcal{C}_1$ be a collection of all cubes from $\mathcal{C}_1$ that contain a point $\widetilde{x} \in \mathcal{H}$ for which there is an $x \in A$ such that $P(x) = \widetilde{x}$ and $h(x) > H/2$.

Each cube from $\mathcal{C}_1 \setminus \mathcal{D}_1$ we split further, such that we get $2^{n-1}$ cubes with side $H/L$. The collection of these cubes we denote by $\mathcal{C}_2$. Now let $\mathcal{D}_2 \subset \mathcal{C}_2$ be all cubes from $\mathcal{C}_2$ containing a point $\widetilde{x} \in \mathcal{H}$, for which there is an $x \in A$ such that $P(x) = \widetilde{x}$ and $h(x) > H/4$. From here we continue inductively. At each stage we split all cubes from $\mathcal{C}_n \setminus \mathcal{D}_n$ into $2^{n-1}$ new cubes with sides half of the previous one. Then we put into $\mathcal{D}_n$ are cubes from $\mathcal{C}_n$ containing a point $\widetilde{x} \in \mathcal{H}$, for which there is an $x \in A$ such that $P(x) = \widetilde{x}$ and $h(x) > H/2^n$.

By $\mathcal{D}$ we denote the union of all $\mathcal{D}_n$, i.e., a cube belongs to $\mathcal{D}$ if and only if it was selected at a certain stage of the process defined above. The set $\mathcal{D}$ is countable and therefore we can order its elements into a sequence $D_1, D_2, D_3, \ldots$. Denote by $C_i$ the preimage of a cube $D_i$ on $\partial\Omega$, i.e.,

$$C_i = \partial\Omega \cap P^{-1}(D_i). \tag{B.23}$$

The collection of all $C_i$ we denote by $\mathcal{C}$.

There are several important observations to make. The first one is,, that the collection $\mathcal{C}$ covers $A$. From this obviously

$$(B.24) \qquad \sigma(A) \leq \sigma(\cup C_i) = \sum_{i=1}^{\infty} \sigma(C_i).$$

Also, since $\partial\Omega$ is Lipschitz, there are positive constants $c_3$ and $c_4$, such that for all $i \in N$

$$(B.25) \qquad c_3 \lambda^{n-1}(D_i) \leq \sigma(C_i) \leq c_4 \lambda^{n-1}(D_i).$$

Here $\lambda^{n-1}$ is the $(n-1)$-dimensional Lebesgue measure on $\mathcal{H}$.

The other comment is that (B.24) can be reversed. Fix $i \in N$. Denote by $r$ the length of the side of $D_i$. From our construction follows that there exists $x \in C_i \cap A$ such that

$$(B.26) \qquad h(x) > r\tfrac{L}{2}.$$

This means that the whole part of downward pointing cone $\widetilde{\gamma}(x', \varphi(x') + r\tfrac{L}{2})$ (here $x = (x', \varphi(x'))$) that lies in $\Omega$ belongs to $\mathcal{A}$, and also the set

$$(B.27) \qquad V = \partial\Omega \cap \widetilde{\gamma}(x', x_n + r\tfrac{L}{2})$$

is a subset of $A$. Now a simple geometric argument gives us that the surface measure of the intersection of $V$ with $C_i$ is at least $Cr^{n-1}$, where the constant $C > 0$ depends only on Lipschitz constant $L$ of $\partial\Omega$. This yields

$$(B.28) \qquad \sigma(A) \geq C\sigma(\cup C_i) = C\sum_{i=1}^{\infty} \sigma(C_i).$$

The estimate (B.28) is crucial. For each $i$ we define a set $E_i$. Let $r_i$ be the length of the side of $D_i$. Let $\widetilde{x}_i$ be the center of the $(n-1)$ dimensional cube $D_i$ in $\mathcal{H}$. We lift this point onto $\partial\Omega$, i.e., we get a point $x_i = (x'_i, \varphi(x'_i)) \in \partial\Omega$ for which $P(x_i) = \widetilde{x}_i$. Finally, we set

$$(B.29) \qquad E_i = \{y = (y', y_n) : |y' - x'_i| \leq Lr_i \text{ and } |y_n - \varphi(x'_i)| \leq Lr_i\},$$

so that $E_i$ is an $n$-dimensional "cube" (naturally just in our coordinates) with center at $x_i$ and side of length $2Lr_i$. This "cube" was carefully picked such that

$$(B.30) \qquad \mathcal{A}_i = \{w \in \mathcal{A} : P(w) \in D_i\} \subset E_i.$$

In particular, the union of all $E_i$ covers $\mathcal{A}$. Finally, we pick a ball $B_i$ with center at $x$ such that $E_i \subset B_i$. Clearly this all can be done such that

$$(B.31) \qquad \text{Vol}(B_i) \approx r_i^n.$$

Define functions $f_i$ by:

$$(B.32) \qquad f_i = f\chi_{\mathcal{A}_i}, \qquad i = 1, 2, 3, \ldots.$$

Here the set $\mathcal{A}_i$ is defined in (B.30) and $\chi_{\mathcal{A}_i}$ is the characteristic function of the set $\mathcal{A}_i$. Obviously,

$$(B.33) \qquad f = \sum_{i=1}^{\infty} f_i.$$

Finally, we put $u_i = L^{-1} f_i$. Since $\mathcal{A}_i \subset B_i$ and $f_i$ satisfies all assumptions of Proposition B.2, we get an $L^p$ estimate on $u_i^*$

(B.34) $$\|u_i^*\|_{L^p(\partial\Omega)} \leq C R_i^{n-1},$$

where $R_i$ is the radius of $B_i$. Since $r_i \approx R_i$ and $r_i^{n-1} \approx \sigma(C_i)$ we get that

(B.35) $$\|u_i^*\|_{L^p(\partial\Omega)} \leq C \sigma(C_i).$$

Finally, we write $u = L^{-1} f$ as $u = \sum u_i$, hence (B.35) and (B.28) give us

(B.36) $$\|u^*\|_{L^p(\partial\Omega)} \leq \sum_{i=1}^{\infty} \|u_i^*\|_{L^p(\partial\Omega)} \leq C \sum_{i=1}^{\infty} \sigma(C_i) \leq C \sigma(A).$$

This concludes our proof. □

Finally, the previous Proposition gives us the following.

THEOREM B.5. *Assume that $f: M \to \mathbb{R}$ is a function with support in $\Omega$ whose nontangential maximal function $f^*$ belongs to $L^1(\partial\Omega)$. Consider the solution to*

(B.37) $$Lu = f \quad \text{on } M, \quad \text{i.e.,} \quad u = L^{-1} f.$$

*For any $1 \leq p < (n-1)/(n-3)$ the nontangential maximal function of $u$; $u^*$ belongs to $L^p(\partial\Omega)$ and there exists a constant $C_p = C(p, M, \Omega) > 0$ such that*

(B.38) $$\|u^*\|_{L^p(\partial\Omega)} \leq C_p \|f^*\|_{L^1(\partial\Omega)}.$$

PROOF. Consider the sets

(B.39) $$A_i = \{x \in \partial\Omega : f^*(x) > i\}.$$

Here, if we want to be completely precise we should consider a partition of unity on $\partial\Omega$ and sets $A_i$ is each coordinate chart corresponding to this partition separately. This is because on two different charts the nontangential approach region $\gamma(x)$ to a point $x \in \partial\Omega$ might slightly differ. This also means that the sets $A_i$ would slightly differ on such two charts. Nevertheless the definition (B.39) is "generically" correct.

Since we took open nontangential approach regions $\gamma(.)$, it follows that each set $A_i$ is open. The fact that $f^* \in L^1(\partial\Omega)$ is equivalent to

(B.40) $$\sum_{i=1}^{\infty} \sigma(A_i) < \infty.$$

Now we write the function $f$ as a infinite sum $\sum f_i$ with functions $f_i$ defined as follows.

(B.41)
$$f_0(x) = \begin{cases} f(x), & \text{if } -1 \leq f(x) \leq 1, \\ 1, & \text{if } 1 < f(x), \\ -1, & \text{if } 1 < -f(x), \end{cases}$$

$$f_i(x) = \begin{cases} 0, & \text{if } |f(x)| \leq i, \\ f(x) - i, & \text{if } i < f(x) \leq i+1, \\ f(x) + i, & \text{if } i < -f(x) \leq i+1, \\ 1, & \text{if } i+1 < f(x), \\ -1, & \text{if } i+1 < -f(x). \end{cases} \quad i = 1, 2, 3, \ldots$$

Notice, that for each $f_i$ we have $|f_i| \leq 1$. There is a connection between the support of each function $f_i$ and the set $A_i$. We claim that

(B.42) $$\operatorname{supp} f_i \subset \operatorname{Pim}(A_i).$$

Seeing this is quite easy. Consider one coordinate chart (B.10). If $x = (x', x_n) \in \operatorname{supp} f_i$ then clearly $|f(x)| > i$. Take any point $z$ from the intersection of $\partial\Omega$ with downward opening cone $\widetilde{\gamma}(x)$. The claim is, that such point is in $A_i$. Indeed, since $x \in \gamma(z)$ we have that $f^*(z) \geq |f(x)| > i$. From this fact that $x \in \operatorname{Pim}(A_i)$ follows immediately.

Now we can proceed. Define $u_i = L^{-1} f_i$, for $i = 0, 1, 2, \ldots$. We use Proposition B.4 to estimate $u_i^*$ for each $i$. This yields

(B.43) $$\|u_i^*\|_{L^p} \leq C_p \sigma(A_i).$$

We estimate $u_0^*$ using the well known fact that given $f_0 \in L^\infty(M)$ $u_0 = L^{-1} f_0$ is a continuous function, hence bounded. This finally gives us:

(B.44) $$\|u^*\|_{L^p} \leq \sum \|u_i^*\|_{L^p} \leq C_p \sum \sigma(A_i) \approx 1 + \int_{\partial\Omega} f^* \, d\sigma = 1 + \|f^*\|_{L^1(\partial\Omega)}.$$

This 'almost' establishes the desired estimate (B.38). The unwanted term '1+' comes from estimating norm of $u_0$. However, a simple scaling argument (using linearity of $L^{-1}$) gives us that $u = \frac{1}{K} L^{-1}(Kf)$, hence $\|u^*\|_{L^p} \leq \frac{C_p}{K}(1 + K\|f^*\|)_{L^1(\partial\Omega)}$. As we let $K \to \infty$ (B.38) follows. $\square$

THEOREM B.6. *Let $\Omega \subset M$ be a domain with $C^1$ boundary. Assume that $f : \Omega \to \mathbb{R}$ is a function whose nontangential maximal function $f^*$ belongs to $L^1(\partial\Omega)$. Let $u$ be the solution to the following Dirichlet problem boundary problem on $\Omega$:*

(B.45) $$Lu = f \quad \text{in } \Omega, \qquad u\big|_{\partial\Omega} = 0, \qquad u^* \in L^1(\partial\Omega).$$

*Then the solution $u$ exists and is unique and moreover, for any $1 \leq p < (n-1)/(n-3)$ the nontangential maximal function of $u$; $u^*$ belongs to $L^p(\partial\Omega)$ and there exists a constant $C_p = C(p, M, \Omega) > 0$ such that*

(B.46) $$\|u^*\|_{L^p(\partial\Omega)} \leq C_p \|f^*\|_{L^1(\partial\Omega)}.$$

If $\Omega$ is a domain with Lipschitz boundary, all above remains true, provided $n \leq 5$.

PROOF. The proof is based on Proposition B.5. Define a function $F$ on $M$ by extending $f$ onto the whole $M$:

$$(B.47) \qquad F(x) = \begin{cases} f(x), & \text{for } x \in \Omega \\ 0, & \text{otherwise.} \end{cases}$$

Let $U = L^{-1}(F)$. On $\Omega$, clearly

$$(B.48) \qquad L U = f \quad \text{and} \quad \|U^*\|_{L^p(\partial\Omega)} \leq C_p \|f^*\|_{L^1(\partial\Omega)}.$$

Consider now the following boundary problem

$$(B.49) \qquad Lw = 0 \text{ on } \Omega, \quad w\big|_{\partial\Omega} = -U\big|_{\partial\Omega} \in L^p(\partial\Omega).$$

(B.49) is solvable for all $1 < p < \infty$, provided $\Omega$ has a $C^1$ boundary (see Theorem 3.1). If $\partial\Omega$ is Lipschitz, (B.49) is solvable for $2 - \varepsilon < p < \infty$ (see [**22**], [**23**] and [**24**]). The solution to (B.49) satisfies the following estimate on $w^*$:

$$(B.50) \qquad \|w^*\|_{L^p(\partial\Omega)} \leq C \|U\big|_{\partial\Omega}\|_{L^p(\partial\Omega)} \leq C \|U^*\|_{L^p(\partial\Omega)} \leq C \|f^*\|_{L^1(\partial\Omega)}.$$

Finally, $u = U + w$ solves (B.45) and the estimate (B.46) follows from (B.48) and (B.50). $\square$

Using interpolation methods we can get more general variant of Theorem B.6. Denote by $\mathcal{D}^{0,p}$ the following set

$$(B.51) \qquad \mathcal{D}^{0,p} = \{f : \Omega \to \mathbb{R}; f^* \in L^p(\partial\Omega)\}, \quad \text{for } 1 \leq p \leq \infty.$$

REMARK. We use the notation $\mathcal{D}^{0,p}$, with first index always zero, because we want to maintain consistency with the papers [**12**] and [**13**], where more general spaces $\mathcal{D}^{s,p}$ are defined.

We claim that $\mathcal{D}^{0,p}$ equipped with the norm

$$(B.52) \qquad \|f\|_{\mathcal{D}^{0,p}} = \|f^*\|_{L^p(\partial\Omega)}$$

is a Banach space for any $p \in [1, \infty]$.

Seeing this is not difficult. Clearly, $\|.\|_{\mathcal{D}^{0,p}}$ satisfies all properties of a norm. We only need to check the completeness. If $(f_n)_{n \in \mathbb{N}}$ is a Cauchy sequence in $\mathcal{D}^{0,p}$, then

$$(B.53) \qquad \|f_n^* - f_m^*\|_{L^p(\partial\Omega)} \leq \|(f_n - f_m)^*\|_{L^p(\partial\Omega)} \to 0, \quad \text{as } m, n \to \infty.$$

Pick any $x \in \Omega$. Consider any $y \in \widetilde{\gamma(x)} \cap \partial\Omega$. We have for such $y$:

$$(B.54) \qquad |f_n(x) - f_m(x)| \leq |(f_n - f_m)^*(y)|,$$

and therefore
$$(B.55)$$
$$|f_n(x) - f_m(x)| \leq C \int_{\widetilde{\gamma(x)} \cap \partial\Omega} |(f_n - f_m)^*(y)| \, d\sigma(y) \leq C \|(f_n - f_m)^*\|_{L^p(\partial\Omega)} \to 0.$$

This means that for any $x$ the sequence $(f_n(x))_{n\in\mathbb{N}}$ is Cauchy, hence convergent. This allows us to define

(B.56) $$f(x) = \lim_{n\to\infty} f_n(x), \quad \text{for } x \in \Omega.$$

Take now any $\varepsilon > 0$ and find $k$, such that for any pair $n, m \geq k$, the difference $(f_n - f_m)^*$ has the $L^p(\partial\Omega)$ norm less than $\varepsilon$. For such $k$ we define functions $g_n$ by

(B.57) $$g_n = (f_n - f_k)^*, \quad n = k, k+1, \ldots.$$

Let $g = \liminf_{n\to\infty} g_n$. By the Fatou's lemma the $L^p$ norm of $g$ is less or equal to $\varepsilon$. For any $y \in \partial\Omega$ and $x \in \gamma(y)$ we have $|f_n(x) - f_k(x)| \leq g_n(y)$. Hence by taking $\liminf_{n\to\infty}$ on both sides of this inequality we get that

(B.58) $$|f(x) - f_k(x)| = \lim_{n\to\infty} |f_n(x) - f_k(x)| \leq \liminf_{n\to\infty} g_n(y) = g(y),$$

This gives

(B.59) $$(f - f_k)^*(y) = \sup_{x\in\gamma(y)} |f(x) - f_k(x)| \leq g(y),$$

and therefore

(B.60) $$\|f - f_k\|_{\mathcal{D}^{0,p}} = \|(f - f_k)^*\|_{L^p(\partial\Omega)} \leq \|g\|_{L^p(\partial\Omega)} < \varepsilon.$$

This proves compleatness.

Our next goal is to study complex interpolation on the spaces $\mathcal{D}^{0,p}$, $1 \leq p \leq \infty$. Recall quickly a simple case of complex interpolation scheme we would like to use.

Let $E$, $F$ be Banach spaces. Suppose that $F$ is included in $E$ and the inclusion $F \hookrightarrow E$ is continuous. If $\mathcal{O}$ is the vertical strip in the complex plane,

(B.61) $$\mathcal{O} = \{z \in \mathbb{C};\ 0 < \operatorname{Re} z < 1\},$$

we define

(B.62) $$\begin{aligned}\mathcal{H}_{E,F}(\mathcal{O}) = \{u(z) &\text{ bounded and continuous on } \overline{\mathcal{O}} \text{ with values in } E;\\ &\text{holomorphic on } \mathcal{O}\colon \|u(1+iy)\|_F \text{ is bounded for } y \in \mathbb{R}\}.\end{aligned}$$

For $\theta \in [0, 1]$ we put

(B.63) $$[E, F]_\theta = \{u(\theta); u \in \mathcal{H}_{E,F}(\mathcal{O})\}.$$

We give $[E, F]_\theta$ the Banach space topology, making it isomorphic to the quotient

(B.64) $$\mathcal{H}_{E,F}(\Omega)/\{u : u(\theta) = 0\}.$$

For convenience we use the convention: $[E, F]_\theta = [F, E]_{1-\theta}$.

PROPOSITION B.7. *For $0 < \theta < 1$ and $1 \leq p_1 < p_2 \leq \infty$,*

(B.65) $$[\mathcal{D}^{0,p_1}, \mathcal{D}^{0,p_2}]_\theta = \mathcal{D}^{0,q},$$

*where $p_1$, $p_2$ and $q$ are related by*

(B.66) $$\frac{1}{q} = \frac{1-\theta}{p_1} + \frac{\theta}{p_2}.$$

PROOF. Given $f \in \mathcal{D}^{0,q}$, we define

(B.67) $$u(z) = |f(x)|^{c(\theta-z)} f(x);$$

$u$ by convention zero when $f(x) = 0$. The number $c$ is chosen such that $u$ belongs to $\mathcal{H}_{\mathcal{D}^{0,p_1}, \mathcal{D}^{0,p_2}}(\mathcal{O})$, which gives $\mathcal{D}^{0,q} \subset [\mathcal{D}^{p_1}, \mathcal{D}^{p_2}]_\theta$. This proves one inclusion.

The other inclusion follows from the following argument. Since $\mathcal{M}^0$ is sublinear and maps $\mathcal{D}^{0,p_j}$ boundedly into $L^{p_j}(\partial\Omega)$ for $j = 1, 2$, a real interpolation gives us that $\mathcal{M}^0$ maps $(\mathcal{D}^{0,p_1}, \mathcal{D}^{0,p_2})_{\theta,\infty}$ into $L^q(\partial\Omega)$, thus, $(\mathcal{D}^{0,p_1}, \mathcal{D}^{0,p_2})_{\theta,\infty} \hookrightarrow \mathcal{D}^{0,q}$. Now, according to well-known connection between the complex and the real methods of interpolation, $[\mathcal{D}^{0,p_1}, \mathcal{D}^{0,p_2}]_\theta \hookrightarrow (\mathcal{D}^{0,p_1}, \mathcal{D}^{0,p_2})_{\theta,\infty}$, hence the inclusion $[\mathcal{D}^{0,p_1}, \mathcal{D}^{0,p_2}]_\theta \subset \mathcal{D}^{0,q}$ follows. □

Next, we establish that for $s > (n-1)/2$ the operator $L^{-1}$ maps $\mathcal{D}^{0,s}$ into $\mathcal{D}^{0,\infty} = L^\infty(\Omega)$.

THEOREM B.8. *Assume that $f : M \to \mathbb{R}$ is a function with support in $\Omega$ and $f \in \mathcal{D}^{0,s}$ for some $s > (n-1)/2$. If $u$ solves*

(B.68) $$Lu = f \quad \text{on } M, \quad \text{i.e.,} \quad u = L^{-1}f,$$

*then $u \in L^\infty(M)$. In particular $u|_\Omega \in L^\infty(\Omega)$. Moreover, there exists a constant $C = C(M,\Omega) > 0$ such that*

(B.69) $$\|u\|_{L^\infty(M)} \leq C \|f^*\|_{\mathcal{D}^{0,s}}.$$

PROOF. The main idea of the proof is very similar to what we did before. Therefore we will be brief. Consider first that $\|f\|_{L^\infty(M)} \leq 1$ and $\operatorname{supp} f \subset \mathcal{A}$. Here $\mathcal{A} = \operatorname{Pim}(A)$ for some $A \subset \partial\Omega$ open. Then we have

(B.70) $$|u(x)| \leq \int_\Omega |E(x,y) f(y)| \, d\operatorname{Vol}(y) \leq \int_\mathcal{A} |E(x,y)| \, d\operatorname{Vol}(y).$$

By (B.2), $|E(x,y)|^q \in L^1(M)$ for any $1 \leq q < n/(n-2)$. Hence, by Hölder inequality we can further estimate (B.70). This gives:

(B.71) $$|u(x)| \leq \left( \int_M |E(x,y)|^q \, d\operatorname{Vol}(y) \right)^{1/q} \left( \int_\mathcal{A} 1 \, d\operatorname{Vol}(y) \right)^{1/p} \leq C(q) \operatorname{Vol}(\mathcal{A})^{1/p}.$$

Here $1/p + 1/q = 1$, which gives that (B.71) is true for any $n/2 < p < \infty$. Finally, if $\mathcal{A} = \operatorname{Pim}(A)$, then

(B.72) $$\operatorname{Vol}(\mathcal{A}) \leq C \sigma(A)^{n/(n-1)}.$$

This inequality follows from the decomposition that has been described in details in the proof of Proposition B.4. We decomposed the set $A$ into disjoint countable union of sets $C_i$ (which were essentially $n-1$ dimensional 'cubes'), such that for each $C_i$ there is a $n$-dimensional ball $B_i$ with the property $\operatorname{diam}(C_i) \approx \operatorname{diam}(B_i)$ and $\mathcal{A} \subset \bigcup B_i$. From this

$$\operatorname{Vol}(\mathcal{A}) \leq C \sum \operatorname{diam}(B_i)^n \approx C \sum \operatorname{diam}(C_i)^n \leq C \sum \sigma(C_i)^{n/(n-1)}$$

(B.73) $$\leq C \sigma \left( \bigcup C_i \right)^{n/(n-1)} = C \sigma(A)^{n/(n-1)}.$$

Combining (B.71) and (B.72) finally yields

(B.74) $$\|u\|_{L^\infty(M)} \leq C(p)\sigma(A)^{n/(np-p)}, \qquad \text{for any } n/2 < p < \infty.$$

Let $s > (n-1)/2$ and $f \in \mathcal{D}^{0,s}$. We can decompose $f$ in a way that resembles (B.41). Let

(B.75)
$$g_0(x) = \begin{cases} f(x), & \text{if } -1 \leq f(x) \leq 1, \\ 1, & \text{if } 1 < f(x), \\ -1, & \text{if } 1 < -f(x), \end{cases}$$

$$g_i(x) = \begin{cases} 0, & \text{if } |f(x)| \leq i^{1/s}, \\ f(x) - i^{1/s}, & \text{if } i^{1/s} < f(x) \leq (i+1)^{1/s}, \\ f(x) + i^{1/s}, & \text{if } i^{1/s} < -f(x) \leq (i+1)^{1/s}, \qquad i = 1, 2, 3, \ldots \\ (i+1)^{1/s} - i^{1/s}, & \text{if } (i+1)^{1/s} < f(x), \\ -(i+1)^{1/s} + i^{1/s}, & \text{if } (i+1)^{1/s} < -f(x). \end{cases}$$

If we put:

(B.76) $$f_0 = g_0, \qquad f_i = \frac{g_i}{(i+1)^{1/s} - i^{1/s}}, \qquad i = 1, 2, \ldots,$$

then for each $f_i$ we have $|f_i| \leq 1$, and

(B.77) $$f = f_0 + \sum_{i=1}^{\infty} \left[(i+1)^{1/s} - i^{1/s}\right] f_i.$$

Moreover,

(B.78) $$\operatorname{supp} f_i \subset \operatorname{Pim}(A_i),$$

where

(B.79) $$A_i = \{x \in \partial\Omega : f^*(x) > i^{1/s}\}, \qquad i = 1, 2, 3 \ldots.$$

Also $\|f^*\|_{L^s(\partial\Omega)}^s \approx \sum \sigma(A_i)$. The estimate (B.74) can be applied to each function $f_i$, $i \geq 1$. Also $(i+1)^{1/s} - i^{1/s} \approx i^{1/s-1}$. This yields

(B.80)
$$\|L^{-1}f\|_{L^\infty(M)} \leq C \sum i^{1/s-1} \sigma(A_i)^{n/(np-p)}$$
$$\leq C \left(\sum i^{\frac{(1-s)p(n-1)}{s(p(n-1)-n)}}\right)^{\frac{p(n-1)-n}{p(n-1)}} \left(\sum \sigma(A_i)\right)^{n/(np-p)}.$$

In the last estimate was obtained by Hölder inequality. The number in the last line of (B.80) is finite, provided

(B.81) $$\frac{(1-s)p(n-1)}{s(p(n-1)-n)} < -1, \qquad \text{i.e.,} \qquad (1-s)p(n-1) < -s(p(n-1)-n).$$

If we simplify (B.81) we get that $p$ should be chosen such that

(B.82) $$p < s\frac{n}{n-1}.$$

Clearly, for $s > (n-1)/2$, it is always possible to find $p$ for which $n/2 < p < s\frac{n}{n-1}$. Then we have a finite estimate on the norm of $\|L^{-1}f\|_{L^\infty(M)}$. □

Now, we interpolate between the results in Theorem B.5 and B.8.

COROLLARY B.9. *Assume that $f : M \to \mathbb{R}$ is a function with support in $\Omega$ whose nontangential maximal function $f^*$ belongs to $L^r(\partial\Omega)$, $1 \leq r \leq \infty$. Consider the solution to*

(B.83) $$Lu = f \quad \text{on } M, \quad \text{i.e.,} \quad u = L^{-1}f.$$

*For any $1 \leq p < \frac{n-1}{(n-1)/r-2}$, provided $r < (n-1)/2$, and $p = \infty$ otherwise, the nontangential maximal function of $u$; $u^*$ belongs to $L^p(\partial\Omega)$ and there exists a constant $C_p = C(p, M, \Omega) > 0$ such that*

(B.84) $$\|u^*\|_{L^p(\partial\Omega)} \leq C_p \|f^*\|_{L^r(\partial\Omega)}.$$

*In particular, for $r < \infty$, $p$ can always be taken such that $p > r$.*

PROOF. We interpolate between

(B.85) $$\begin{aligned} L^{-1} &: \mathcal{D}^{0,1} \to \mathcal{D}^{0,q} \quad \text{with } q \lesssim \frac{n-1}{n-3}, \\ L^{-1} &: \mathcal{D}^{0,s} \to \mathcal{D}^{0,\infty} \quad \text{with } s \gtrsim \frac{n-1}{2}. \end{aligned}$$

It follows, that we have a sequence of continuous maps

(B.86) $$\mathcal{D}^{0,r} \xhookrightarrow{id} [\mathcal{D}^{0,1}, \mathcal{D}^{0,s}]_\theta \xrightarrow{L^{-1}} [\mathcal{D}^{0,q}, \mathcal{D}^{0,\infty}]_\theta \xhookrightarrow{id} \mathcal{D}^{0,\tilde{r}},$$

for any $\tilde{r} \lesssim \frac{n-1}{(n-1)/r-2}$. □

Corollary B.9 we can be used to substantially improve Theorem B.6.

THEOREM B.10. *Let $\Omega \subset M$ be a domain with $C^1$ boundary. Assume that $f : \Omega \to \mathbb{R}$ is a function whose nontangential maximal function $f^*$ belongs to $L^p(\partial\Omega)$, for some $1 \leq p \leq \infty$. Then the Dirichlet problem*

(B.87) $$Lu = f \quad \text{in } \Omega, \quad u|_{\partial\Omega} = 0, \quad u^* \in L^1(\partial\Omega),$$

*has a unique solution $u$ whose nontangential maximal function $u^*$ belongs to $L^q(\partial\Omega)$ for any $1 \leq q < \frac{n-1}{(n-1)/p-2}$, provided $p < (n-1)/2$, $q = \infty$ otherwise. In particular, always $q \geq p$. Moreover, there exists a constant $C = C(p, q, M, \Omega) > 0$ such that*

(B.88) $$\|u^*\|_{L^q(\partial\Omega)} \leq C \|f^*\|_{L^p(\partial\Omega)}.$$

*If $\Omega$ is a domain with Lipschitz boundary, all above remains true for $\dim M = n \leq 5$. If $n > 5$, (B.87) is solvable for any $\frac{2(n-1)}{n+3} - \varepsilon < p \leq \infty$ where $\varepsilon > 0$ depends on the given domain $\Omega$. Also the estimate (B.88) remains true.*

PROOF. Basically, all said in the proof of Theorem B.6. can be used. Only questionable case is $q = \infty$, but in such situation we apply Proposition 5.7 in [**22**]. □

# Bibliography

[1]  J. Bergh and J.Löfström, Interpolation spaces, Springer-Verlag, 1976.
[2]  A. P. Calderón, *Cauchy integrals on Lipschitz curves and related operators*, Proc. Nat. Acad. Sci. **74** (1977), 1324–1327.
[3]  A. P. Calderón, C. P. Calderón, E. Fabes, M. Jodeit and N. M. Riviere, *Applications of the Cauchy integral on Lipschitz curves*, Bull. Am. Math. Sci. **84** (1978), 287–290.
[4]  R. Coifman, G. David and Y. Meyer, *La solution des conjectures des Calderón*, Adv. Math. **48** (1983), 144–148.
[5]  R. R. Coifman, A. McIntosh and Y. Meyer, *L'integrale de Cauchy définit un opérateur borné sur $L^2$ pour les courbes lipschitziennes*, Ann. Math. **116** (1982), 361–387.
[6]  B. Dahlberg, *On the Poisson integral for Lipschitz and $C^1$ domains*, Studia Math. LXVI. (1979), 13–26.
[7]  B. Dahlberg, *Harmonic functions in Lipschitz domains*, Proceedings of Symposia in Pure Mathematics **XXXV** (1979), 313–322.
[8]  Guy David, *Opérateurs intégraux singuliers sur certaines courbes du plan complexe*, Ann. scient. Éc. Norm. Sup. **17** (1984), 157–187.
[9]  G. David and J. Journé, *A boundedness criterion for generalized Calderón-Zygmund operators*, Ann. Math. **120** (1984), 371–397.
[10] B. Dahlberg and C. Kenig, *Hardy spaces and the Neumann problem in $L^p$ for Laplace's equation in Lipschitz domains*, Ann. Math. **125** (1987), 437–465.
[11] B. Dahlberg, C. Kenig and G. Verchota, *Boundary value problems for the system of elastostatics on Lipschitz domains*, Duke Math. J. **57** (1988), 795–818.
[12] M. Dindoš, *Existence and uniqueness for a semilinear elliptic problem on Lipschitz domains in Riemannian manifolds*, Comm. PDE **27** (2002), 219–291.
[13] M. Dindoš, *Existence and uniqueness for a semilinear elliptic problem on Lipschitz domains in Riemannian manifolds II*, Transactions of AMS **355** (2003), no. 4, 1365–1399.
[14] E. Fabes, M. Jodeit jr. and N. Rivère, *Potential techniques for boundary value problems on $C^1$-domains*, Acta Math. **141** (1978), 165–186.
[15] E. Fabes and C. Kenig, *On Hardy space $H^1$ of a $C^1$ domain*, Arkiv för matematik **19** (1981), 1–22.
[16] E. Fabes, C. Kenig and G. Verchota, *The Dirichlet problem for the Stokes system on Lipschitz domains*, Duke Math. J. **57** (1988), 769–793.
[17] C. Fefferman and E. Stein, *$H^p$ spaces of several variables*, Acta Math. **129** (1972), 137–193.
[18] Carlos Kenig, Harmonic analysis: Techniques for second order elliptic boundary value problems, American Math. Society, 1994.
[19] Carlos Kenig, *Elliptic boundary value problems on Lipschitz domains*, Beijing lectures in Harmonic analysis (E. Stein, ed.), Princeton University Press, 1986, pp. 131–183.
[20] D. Mitrea, M. Mitrea and J. Pipher, *Vector potential theory on nonsmooth domains in $R^3$ and applications to electromagnetic scattering*, J. Fourier Anal. and Appl. **3** (1997), 131–192.
[21] D. Mitrea, M. Mitrea and M. Taylor, *Layer potentials, the Hodge Laplacian, and global boundary problems in nonsmooth Riemannian manifolds*, Memoirs AMS (2001), no. 713.
[22] M. Mitrea and M. Taylor, *Boundary layer methods for Lipschitz domains in Riemannian manifolds*, J. of Funct. Anal. (1999), no. 163, 181–251.
[23] M. Mitrea and M. Taylor, *Potential theory on Lipschitz domains in Riemannian manifolds: $L^p$, Hardy and Hölder space results*, Comm. in Anal. and Geom. **9** (2001), no. 2, 369–421.
[24] M. Mitrea and M. Taylor, *Potential theory on Lipschitz domains in Riemannian manifolds: Hölder continuous metric tensors*, Comm. PDE **25** (2000), 1487–1536.

[25] M. Mitrea and M. Taylor, *Potential theory on Lipschitz domains in Riemannian manifolds: Sobolev-Besov space results nad the Poisson problem*, J. Funct. Anal. (2000), no. 176, 1–79.

[26] E. Stein, Harmonic analysis: Real-Variable methods, Orthogonality and Oscillatory integrals, Princeton University Press, 1993.

[27] E. Stein, Singular integrals and Differentiability Properties of Functions, Princeton University Press, 1970.

[28] E. Stein and G. Weiss, *On the theory of harmonic functions of several variables I, The theory of $H^p$ soaces*, Acta Math. **103** (1960), 25–62.

[29] Michael Taylor, Partial differential equations, vol. 1, Springer-Verlag, 1996.

[30] G. Verchota, *Layer potentials and regularity for the Dirichlet problem for Laplace's operator in Lipschitz domains*, J. Funct. Anal. **59** (1984), 572–611.

[31] J. M. Wilson, *A simple proof of the atomic decomposition for $H^p(R^n)$, $0 < p \leq 1$*, Studia Mathematica LXXIV. (1982).

## Editorial Information

To be published in the *Memoirs*, a paper must be correct, new, nontrivial, and significant. Further, it must be well written and of interest to a substantial number of mathematicians. Piecemeal results, such as an inconclusive step toward an unproved major theorem or a minor variation on a known result, are in general not acceptable for publication.

Papers appearing in *Memoirs* are generally at least 80 and not more than 200 published pages in length. Papers less than 80 or more than 200 published pages require the approval of the Managing Editor of the Transactions/Memoirs Editorial Board.

As of September 30, 2007, the backlog for this journal was approximately 14 volumes. This estimate is the result of dividing the number of manuscripts for this journal in the Providence office that have not yet gone to the printer on the above date by the average number of monographs per volume over the previous twelve months, reduced by the number of volumes published in four months (the time necessary for preparing a volume for the printer). (There are 6 volumes per year, each usually containing at least 4 numbers.)

A Consent to Publish and Copyright Agreement is required before a paper will be published in the *Memoirs*. After a paper is accepted for publication, the Providence office will send a Consent to Publish and Copyright Agreement to all authors of the paper. By submitting a paper to the *Memoirs*, authors certify that the results have not been submitted to nor are they under consideration for publication by another journal, conference proceedings, or similar publication.

## Information for Authors

*Memoirs* are printed from camera copy fully prepared by the author. This means that the finished book will look exactly like the copy submitted.

**Initial submission.** The AMS uses Centralized Manuscript Processing for initial submissions. Authors should submit a PDF file using the Initial Manuscript Submission form found at www.ams.org/cgi-bin/peertrack/submission.pl, or send one copy of the manuscript to the following address: Centralized Manuscript Processing, MEMOIRS OF THE AMS, 201 Charles Street, Providence, RI 02904-2294 USA. If a paper copy is being forwarded to the AMS, indicate that it is for it Memoirs and include the name of the corresponding author, contact information such as email address or mailing address, and the name of an appropriate Editor to review the paper (see the list of Editors below).

The paper must contain a *descriptive title* and an *abstract* that summarizes the article in language suitable for workers in the general field (algebra, analysis, etc.). The *descriptive title* should be short, but informative; useless or vague phrases such as "some remarks about" or "concerning" should be avoided. The *abstract* should be at least one complete sentence, and at most 300 words. Included with the footnotes to the paper should be the 2000 *Mathematics Subject Classification* representing the primary and secondary subjects of the article. The classifications are accessible from www.ams.org/msc/. The list of classifications is also available in print starting with the 1999 annual index of *Mathematical Reviews*. The Mathematics Subject Classification footnote may be followed by a list of *key words and phrases* describing the subject matter of the article and taken from it. Journal abbreviations used in bibliographies are listed in the latest *Mathematical Reviews* annual index. The series abbreviations are also accessible from www.ams.org/publications/. To help in preparing and verifying references, the AMS offers MR Lookup, a Reference Tool for Linking, at www.ams.org/mrlookup/.

**Electronically prepared manuscripts.** The AMS encourages electronically prepared manuscripts, with a strong preference for $\mathcal{A}_\mathcal{M}\mathcal{S}$-LATEX. To this end, the Society has prepared $\mathcal{A}_\mathcal{M}\mathcal{S}$-LATEX author packages for each AMS publication. Author packages include instructions for preparing electronic manuscripts, samples, and a style file that generates the particular design specifications of that publication series. Though $\mathcal{AMS}$-LaTeX is the highly preferred format of TeX, author packages are also available in $\mathcal{AMS}$-TeX.

Authors may retrieve an author package from the AMS website starting from www.ams.org/tex/ or via FTP to ftp.ams.org (login as anonymous, enter username as password, and type cd pub/author-info). The *AMS Author Handbook* and the *Instruction Manual* are available in PDF format following the author packages link from www.ams.org/tex/. The author package can also be obtained free of charge by sending email to tech-support@ams.org (Internet) or from the Publication Division, American Mathematical Society, 201 Charles St., Providence, RI 02904-2294, USA. When requesting an author package, please specify $\mathcal{AMS}$-LaTeX or $\mathcal{AMS}$-TeX and the publication in which your paper will appear. Please be sure to include your complete mailing address.

**After acceptance.** The final version of the electronic file should be sent to the Providence office (this includes any TeX source file, any graphics files, and the DVI or PostScript file) immediately after the paper has been accepted for publication.

Before sending the source file, be sure you have proofread your paper carefully. The files you send must be the EXACT files used to generate the proof copy that was accepted for publication. For all publications, authors are required to send a printed copy of their paper, which exactly matches the copy approved for publication, along with any graphics that will appear in the paper.

Accepted electronically prepared files can be submitted via the web at www.ams.org/submit-book-journal/, sent via FTP, or sent on CD-Rom or diskette to the Electronic Prepress Department, American Mathematical Society, 201 Charles Street, Providence, RI 02904-2294 USA. TeX source files, DVI files, and PostScript files can be transferred over the Internet by FTP to the Internet node ftp.ams.org (130.44.1.100). When sending a manuscript electronically via CD-Rom or diskette, please be sure to include a message identifying the paper as a Memoir.

Electronically prepared manuscripts can also be sent via email to pub-submit@ams.org (Internet). In order to send files via email, they must be encoded properly. (DVI files are binary and PostScript files tend to be very large.)

**Electronic graphics.** Comprehensive instructions on preparing graphics are available at www.ams.org/jourhtml/. A few of the major requirements are given here.

Submit files for graphics as EPS (Encapsulated PostScript) files. This includes graphics originated via a graphics application as well as scanned photographs or other computer-generated images. If this is not possible, TIFF files are acceptable as long as they can be opened in Adobe Photoshop or Illustrator. No matter what method was used to produce the graphic, it is necessary to provide a paper copy to the AMS.

Authors using graphics packages for the creation of electronic art should also avoid the use of any lines thinner than 0.5 points in width. Many graphics packages allow the user to specify a "hairline" for a very thin line. Hairlines often look acceptable when proofed on a typical laser printer. However, when produced on a high-resolution laser imagesetter, hairlines become nearly invisible and will be lost entirely in the final printing process.

Screens should be set to values between 15% and 85%. Screens which fall outside of this range are too light or too dark to print correctly. Variations of screens within a graphic should be no less than 10%.

**Inquiries.** Any inquiries concerning a paper that has been accepted for publication should be sent to memo-query@ams.org or directly to the Electronic Prepress Department, American Mathematical Society, 201 Charles St., Providence, RI 02904-2294 USA.

# Editors

This journal is designed particularly for long research papers, normally at least 80 pages in length, and groups of cognate papers in pure and applied mathematics. Papers intended for publication in the *Memoirs* should be addressed to one of the following editors. The AMS uses Centralized Manuscript Processing for initial submissions to AMS journals. Authors should follow instructions listed on the Initial Submission page found at www.ams.org/memo/memosubmit.html.

**Algebra** to ALEXANDER KLESHCHEV, Department of Mathematics, University of Oregon, Eugene, OR 97403-1222; email: ams@noether.uoregon.edu

**Algebraic geometry and its application** to MINA TEICHER, Emmy Noether Research Institute for Mathematics, Bar-Ilan University, Ramat-Gan 52900, Israel; email: teicher@macs.biu.ac.il

**Algebraic geometry** to DAN ABRAMOVICH, Department of Mathematics, Brown University, Box 1917, Providence, RI 02912; email: amsedit@math.brown.edu

**Algebraic number theory** to V. KUMAR MURTY, Department of Mathematics, University of Toronto, 100 St. George Street, Toronto, ON M5S 1A1, Canada; email: murty@math.toronto.edu

**Algebraic topology** to ALEJANDRO ADEM, Department of Mathematics, University of British Columbia, Room 121, 1984 Mathematics Road, Vancouver, British Columbia, Canada V6T 1Z2; email: adem@math.ubc.ca

**Combinatorics** to JOHN R. STEMBRIDGE, Department of Mathematics, University of Michigan, Ann Arbor, Michigan 48109-1109; email: FRS@umich.edu

**Complex analysis and harmonic analysis** to ALEXANDER NAGEL, Department of Mathematics, University of Wisconsin, 480 Lincoln Drive, Madison, WI 53706-1313; email: nagel@math.wisc.edu

**Differential geometry and global analysis** to LISA C. JEFFREY, Department of Mathematics, University of Toronto, 100 St. George St., Toronto, ON Canada M5S 3G3; email: jeffrey@math.toronto.edu

**Dynamical systems and ergodic theory** to AMIE WILKINSON, Department of Mathematics, Northwestern University, 2033 Sheridan Road, Evanston, IL 60208-2730; email: transactions@math.northwestern.edu

**Functional analysis and operator algebras** to DIMITRI SHLYAKHTENKO, Department of Mathematics, University of California, Los Angeles, CA 90095; email: shlyakht@math.ucla.edu

**Geometric analysis** to WILLIAM P. MINICOZZI II, Department of Mathematics, Johns Hopkins University, 3400 N. Charles St., Baltimore, MD 21218; email: trans@math.jhu.edu

**Geometric analysis** to MLADEN BESTVINA, Department of Mathematics, University of Utah, 155 South 1400 East, JWB 233, Salt Lake City, Utah 84112-0090; email: bestvina@math.utah.edu

**Harmonic analysis, representation theory, and Lie theory** to ROBERT J. STANTON, Department of Mathematics, The Ohio State University, 231 West 18th Avenue, Columbus, OH 43210-1174; email: stanton@math.ohio-state.edu

**Logic** to STEFFEN LEMPP, Department of Mathematics, University of Wisconsin, 480 Lincoln Drive, Madison, Wisconsin 53706-1388; email: lempp@math.wisc.edu

**Partial differential equations** to GUSTAVO PONCE, Department of Mathematics, South Hall, Room 6607, University of California, Santa Barbara, CA 93106; email: ponce@math.ucsb.edu

**Partial differential equations and dynamical systems** to PETER POLACIK, School of Mathematics, University of Minnesota, Minneapolis, MN 55455; email: polacik@math.umn.edu

**Probability and statistics** to KRZYSZTOF BURDZY, Department of Mathematics, University of Washington, Box 354350, Seattle, Washington 98195-4350; email: burdzy@math.washington.edu

**Real analysis and partial differential equations** to DANIEL TATARU, Department of Mathematics, University of California, Berkeley, Berkeley, CA 94720; email: tataru@math.berkeley.edu

**All other communications to the editors** should be addressed to the Managing Editor, ROBERT GURALNICK, Department of Mathematics, University of Southern California, Los Angeles, CA 90089-1113; email: guralnic@math.usc.edu.

# Titles in This Series

895 **Steffen Roch,** Finite sections of band-dominated operators, 2008

894 **Martin Dindoš,** Hardy spaces and potential theory on $C^1$ domains in Riemannian manifolds, 2008

893 **Tadeusz Iwaniec and Gaven Martin,** The Beltrami Equation, 2008

892 **Jim Agler, John Harland, and Benjamin J. Raphael,** Classical function theory, operator dilation theory, and machine computation on multiply-connected domains, 2008

891 **John H. Hubbard and Peter Papadopol,** Newton's method applied to two quadratic equations in $\mathbb{C}^2$ viewed as a global dynamical system, 2008

890 **Steven Dale Cutkosky,** Toroidalization of dominant morphisms of 3-folds, 2007

889 **Michael Sever,** Distribution solutions of nonlinear systems of conservation laws, 2007

888 **Roger Chalkley,** Basic global relative invariants for nonlinear differential equations, 2007

887 **Charlotte Wahl,** Noncommutative Maslov index and eta-forms, 2007

886 **Robert M. Guralnick and John Shareshian,** Symmetric and alternating groups as monodromy groups of Riemann surfaces I: Generic covers and covers with many branch points, 2007

885 **Jae Choon Cha,** The structure of the rational concordance group of knots, 2007

884 **Dan Haran, Moshe Jarden, and Florian Pop,** Projective group structures as absolute Galois structures with block approximation, 2007

883 **Apostolos Beligiannis and Idun Reiten,** Homological and homotopical aspects of torsion theories, 2007

882 **Lars Inge Hedberg and Yuri Netrusov,** An axiomatic approach to function spaces, spec tral synthesis and Luzin approximation, 2007

881 **Tao Mei,** Operator valued Hardy spaces, 2007

880 **Bruce C. Berndt, Geumlan Choi, Youn-Seo Choi, Heekyoung Hahn, Boon Pin Yeap, Ae Ja Yee, Hamza Yesilyurt, and Jinhee Yi,** Ramanujan's forty identities for Rogers-Ramanujan functions, 2007

879 **O. García-Prada, P. B. Gothen, and V. Muñoz,** Betti numbers of the moduli space of rank 3 parabolic Higgs bundles, 2007

878 **Alessandra Celletti and Luigi Chierchia,** KAM stability and celestial mechanics, 2007

877 **María J. Carro, José A. Raposo, and Javier Soria,** Recent developments in the theory of Lorentz spaces and weighted inequalities, 2007

876 **Gabriel Debs and Jean Saint Raymond,** Borel liftings of Borel sets: Some decidable and undecidable statements, 2007

875 **C. Krattenthaler and T. Rivoal,** Hypergéométrie et fonction zêta de Riemann, 2007

874 **Sonia Natale,** Semisolvability of semisimple Hopf algebras of low dimension, 2007

873 **A. J. Duncan,** Exponential genus problems in one-relator products of groups, 2007

872 **Anthony V. Geramita, Tadahito Harima, Juan C. Migliore, and Yong Su Shin,** The Hilbert function of a level algebra, 2007

871 **Pascal Auscher,** On necessary and sufficient conditions for $L^p$-estimates of Riesz transforms associated to elliptic operators on $\mathbb{R}^n$ and related estimates, 2007

870 **Takuro Mochizuki,** Asymptotic behaviour of tame harmonic bundles and an application to pure twistor $D$-modules, Part 2, 2007

869 **Takuro Mochizuki,** Asymptotic behaviour of tame harmonic bundles and an application to pure twistor $D$-modules, Part 1, 2007

868 **Gelu Popescu,** Entropy and multivariable interpolation, 2006

867 **Vilmos Totik,** Metric properties of harmonic measures, 2006

866 **William Craig,** Semigroups underlying first-order logic, 2006

865 **Nathanial P. Brown,** Invariant means and finite representation theory of $C*$-algebras, 2006

## TITLES IN THIS SERIES

864 **John M. Lee,** Fredholm operators and Einstein metrics on conformally compact manifolds, 2006

863 **M. Lübke and A. Teleman,** The Universal Kobayashi-Hitchin correspondence on Hermitian manifolds, 2006

862 **Alberto Canonaco,** The Beilinson complex and canonical rings of irregular surfaces, 2006

861 **Leon A. Takhtajan and Lee-Peng Teo,** Weil-Petersson metric on the universal Teichmüller space, 2006

860 **Thomas M. Fiore,** Pseudo limits, biadjoints and pseudo algebras: Categorical foundations of conformal field theory, 2006

859 **N. Arcozzi, R. Rochberg, and E. Sawyer,** Carleson measures and interpolating sequences for Besov spaces on complex balls, 2006

858 **Enrico Valdinoci, Berardino Sciunzi, and Vasile Ovidiu Savin,** Flat level set regularity of $p$-Laplace phase transitions, 2006

857 **Donatella Danielli, Nocola Garofalo, and Duy-Minh Nhieu,** Non-doubling Ahlfors measures, perimeter measures, and the characterization of the trace spaces of Sobolev functions in Carnot-Carathéodory spaces, 2006

856 **Vladimir Bolotnikov and Harry Dym,** On boundary interpolation for matrix valued Schur functions, 2006

855 **Yevgenia Kashina, Yorck Sommerhäuser, and Yongchang Zhu,** On higher Frobenius-Schur indicators, 2006

854 **Noam Greenberg,** The role of true finiteness in the admissible recursively enumerable degrees, 2006

853 **Joachim Krieger,** Stability of spherically symmetric wave maps, 2006

852 **Viorel Barbu, Irena Lasiecka, and Roberto Triggiani,** Tangential boundary stabilization of Navier-Stokes equations, 2006

851 **Jie Wu,** On maps from loop suspensions to loop spaces and the shuffle relations on the Cohen groups, 2006

850 **Siegfried Echterhoff, S. Kaliszewski, John Quigg, and Iain Raeburn,** A categorical approach to imprimitivity theorems for $C^*$-dynamical systems, 2006

849 **Katsuhiko Kuribayashi, Mamoru Mimura, and Tetsu Nishimoto,** Twisted tensor products related to the cohomology of the classifying spaces of loop groups, 2006

848 **Bob Oliver,** Equivalences of classifying spaces completed at the prime two, 2006

847 **Eric T. Sawyer and Richard L. Wheeden,** Hölder continuity of weak solutions to subelliptic equations with rough coefficients, 2006

846 **Victor Beresnevich, Detta Dickinson, and Sanju Velani,** Measure theoretic laws for lim–sup sets, 2006

845 **Ehud Friedgut, Vojtech Rödl, Andrzej Ruciński, and Prasad V. Tetali,** A Sharp threshold for random graphs with a monochromatic triangle in every edge coloring, 2006

844 **Amadeu Delshams, Rafael de la Llave, and Tere M. Seara,** A geometric mechanism for diffusion in Hamiltonian systems overcoming the large gap problem: Heuristics and rigorous verification on a model, 2006

843 **Denis V. Osin,** Relatively hyperbolic groups: Intrinsic geometry, algebraic properties, and algorithmic problems, 2006

842 **David P. Blecher and Vrej Zarikian,** The calculus of one-sided $M$-ideals and multipliers in operator spaces, 2006

For a complete list of titles in this series, visit the
AMS Bookstore at **www.ams.org/bookstore/**.